MINERALS & ROCKS

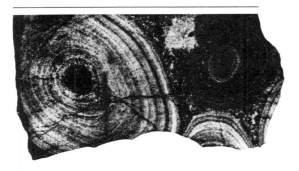

Brian Simpson

M.Sc., (Liverpool), F.G.S.,
Companion of the Institute of Civil Engineers
Professor of Field Studies,
New England College, Arundel

MINERALS & ROCKS

Octopus Books

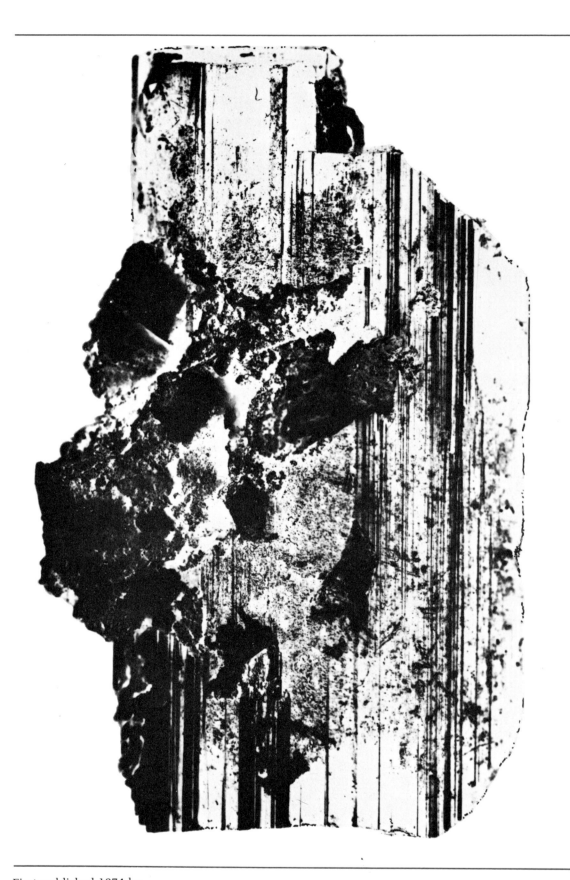

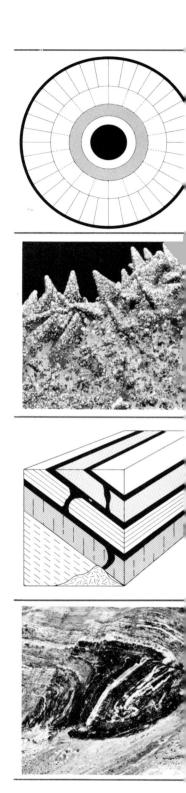

First published 1974 by
Octopus Books Limited
59 Grosvenor Street, London W1

ISBN 0 7064 0358 4

© 1974 Octopus Books Limited

Produced by Mandarin Publishers Limited
14 Westlands Road, Quarry Bay, Hong Kong
Printed in Hong Kong

CONTENTS

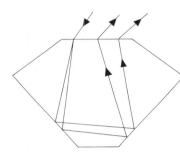

INTRODUCTION

The study of minerals and rocks forms one small part of the science of *geology* – the science concerned with the study of the earth as a whole. Geology, like any other science, can be broken down into a number of specialized topics. Research in each of these topics provides facts which, when co-ordinated, help to build up a complete picture of the nature and past history of the earth. The specialized topics are mineralogy, crystallography, petrology, palaeontology, stratigraphy, structural geology, geochemistry, geophysics, and applied and economic geology. The flow chart on page 9 shows how these topics are inter-related and how they help to build up an overall picture of our planet.

Mineralogy is the study of minerals, and a mineral is an inorganic (non-animal or non-vegetable) substance. There are more than one hundred elements known to science, and each of them has its own particular set of properties. The smallest particles of matter which retain these properties occur as atoms of the elements. The atoms act as building blocks and join together in various ways to form minerals. Minerals, in turn, are the units from which the rocks are built, for example, atoms of silicon and oxygen combine together to form the mineral quartz – which is the main constituent of the sand found on our beaches; and atoms of chlorine and sodium join together to form the mineral halite – salt. It is not necessary to be a skilled scientist to enjoy the science of mineralogy, for the collection and study of minerals provides an interesting and rewarding hobby for the amateur geologist. This book explains how to recognize and collect minerals, and gives simple scientific explanations of their formation and industrial uses.

Some of the rocks you see may be composed of several different minerals locked together in a mass. Other minerals may be found isolated and perfectly formed in regular three-dimensional patterns. Such mineral formations are known as crystals, and the study of crystal formation is known as *crystallography*. Investigation has shown that the atoms forming the minerals are themselves arranged in geometric patterns. For example, salt crystals occur in the form of a cube, and emerald crystals occur in hexagonal (six-sided) columns. The regularity of crystal formation can be demonstrated by means of X-ray photographs. The X-rays pass through the crystal and are deflected by the individual atoms. This produces a pattern of dots on a photographic plate. From this pattern the symmetrical arrangement of the atoms can be seen and measured.

The Matterhorn, Switzerland Spectacular evidence exists in many parts of the world of the mountain building processes which have occured throughout geological history. The Matterhorn is, geologically speaking, a relatively young mountain, having been formed only some 50,000,000 years ago.

GEOLOGICAL TIME SCALE   *Note: each division equals 10 million years

Age*	Era	Period		Dominant Life in the Seas	Dominant Life on Land	Main Mountains Formed
	CAINOZOIC	Neogene	TERTIARY	lamellibranchs and gastropods	mammals and flowering plants	Rocky Mountains Andes Mountains Alps Himalayan Mountains
100		Paleogene				
				ALPINE		
	MESOZOIC	Cretaceous		ammonites and belemnites	dinosaurs and conifer trees	
		Jurassic				
200		Triassic				
	PALAEOZOIC — UPPER	Permian		corals and brachiopods	early amphibian and reptiles and fern-like plants	Appalachian Mountains Ural Mountains
300		Carboniferous				
		Devonian		primitive fish		Scandinavian Mountains Scottish & Welsh Mountains Acadian Mountains
400		Silurian		trilobites	primitive land plants	
	PALAEOZOIC — LOWER	Ordovician		graptolites and crinoids	no life on land	
500		Cambrian				
600	CRYPTOZOIC	Pre-Cambrian		jelly fish and algae		

The study of rocks is known as *petrology*, from the Greek word *petra*, which means *rock*. (The name *Peter* is derived from the same source.) The rocks which compose the earth's outer layer are classified according to the ways in which they were formed, and the three main groups are igneous rocks, sedimentary rocks and metamorphic rocks. Igneous rocks are formed when hot fluid rock cools and solidifies. A wide variety of igneous rocks occur. They vary in mineral content, colour, and in the size of their constituent crystals. The mineral content is determined by the composition of the magma – the molten body of rock substance – from which the rock was formed. One type of magma is rich in silica and alumina and is known as *acid magma*. Such magma produces light-coloured rocks. Another is less rich in silica, but contains quantities of iron and magnesium. This is known as *basic magma* and produces dark-coloured rocks. Igneous rocks can also be classified according to the location in which they solidified. One group solidifies on the earth's surface at normal temperatures and pressures – volcanic lavas, for example. Rocks belonging to this group are known as *extrusive* igneous rocks. They cool relatively quickly, and the individual crystals have little time to develop. Thus small crystals are a feature of this group. The second group cools and solidifies within the earth, below the surface. Such rocks are known as *intrusive* igneous rocks. Because of the higher temperature and greater pressures found within the earth, the cooling process is usually much slower. Crystals have time to develop and so intrusive igneous rocks generally contain larger crystals. The granites which form the cores of great mountain chains such as the Andes and the Rocky Mountains are intrusive igneous rocks; the polygonal columns of basalt found in New Zealand and Northern Ireland are examples of

<table>
<tr><td colspan="3" align="center">**The Surface of the Earth**</td></tr>
</table>

Stratigraphy			Atoms
			↓
Mineralogy/Crystallography }		**Minerals**	

Petrology

Igneous Rocks eg. granite basalt	Metamorphic Rocks eg. slate marble	Sedimentary Rocks eg. chalk sandstone	**Rocks**

Applied and Economic Geology

↑ ↑↑ ↑

Structural Geology

	Heat and Pressure	worn down by weathering and erosion (landscaping)
Molten rock (magma)		
	deposition ← sediments	
	Palaeontology } fossils	
		animals and plants

Geochemistry Geophysics }	**The Interior of the Earth**

OPPOSITE **Table showing the geological time scale** The periods of major mountain building are related to the dominant life in the seas and on land.

LEFT **Table showing the inter-relationship between the various branches of geology**

BELOW LEFT **Section through the Earth** A *The crust* This is divided into an upper layer (sial) of granitic composition and a lower layer (sima) of basaltic composition. B *Mohorovičić Discontinuity* C *The upper mantle* Peridotite D *The lower mantle* Peridotite E *The Oldham-Gutenburg Discontinuity* F *The outer core* G *The transition zone. The inner core.* The outer core, transition zone and inner core are of nickel/iron. Thickness and radius in kilometres (after Holmes): B – 33 km, C – 984 km, E – 2898 km, F – 4703 km, G – 5154 km, centre 6371 km.

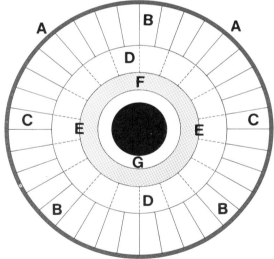

Examples of three rock types:

LEFT **Quiberon, Brittany** Most coastlines have well exposed rock sections; here the hard igneous rock granite forms a formidable barrier to the sea.

ABOVE LEFT **Juan River valley, Cortez, Colorado** The meandering river has cut deeply into sedimentary rock. The strata can be seen as horizontal layers in the sides of the valley.

FAR LEFT **Burra Voe, Shetland** An outcrop of the metamorphic rock gneiss, clearly showing the alternately banded character of the rock. The white bands are composed of feldspar and quartz. The darker layers are composed of hornblende and ferromagnesian minerals.

extrusive igneous rocks. Breakdown processes are described as 'weathering'. This phenomenon consists of two processes – disintegration or mechanical weathering, and decomposition or chemical weathering. The former process produces particles of minerals and rocks, whilst in the latter process, mineral matter is taken into solution and new minerals may be formed. Mechanical weathering results mostly from temperature changes: at the extremes are the expansion and contraction of rocks in hot deserts where great daily temperature variation occurs, and the wedging effect of the freezing-thawing cycle in Arctic regions. The latter form calcite and feldspar, for example, or sodium from the feldspars; new minerals, such as the clay minerals, are formed by the hydrolysis of feldspars.

The products of weathering are usually removed from their place of origin by erosion – through the action of water, wind and gravity – are transported to an area such as the sea, and there they are deposited. They are finally consolidated to form a sedimentary rock.

Some products of disintegration become sands and finally sandstones. The finer products of disintegration and the new clay minerals become shales and mudstones. Dissolved substances, such as lime, may be precipitated as limestones whilst sodium through evaporation of the water, may accumulate as salt.

The process of breakdown, transportation, accumulation and formation is a continuous one. The sand on our beaches and in our deserts, and the pebbles and mud in our streams and rivers, are the fragments which will eventually become sedimentary rocks.

The squeezing and heating of existing rocks, or the introduction of chemically active fluids or gases, results in the formation of a third group of rocks; for

9

example, if a body of magma is brought into contact with a limestone, the limestone layer recrystallizes. If the limestone is pure, recrystallization results in the formation of marble. If a sequence of muddy rocks is subjected to great pressure over long periods of time, the muddy rocks will be transformed into slate. This process of change is known as *metamorphism*, and the resulting rocks as *metamorphic* rocks. These three groups of rocks – igneous, sedimentary and metamorphic – form the substance of the study of petrology, and each group is discussed more fully in subsequent chapters.

The study of sedimentary rocks is particularly interesting for it is within this group of rocks that the remains of ancient plants and animals are to be found. These former organisms have been preserved in various ways, and are known as fossils. The study of fossils is known as *palaeontology*. This rather complicated word is also derived from the Greek. *Palae* means 'old', *onto*

means 'life' and *ology* means 'study', so that palaeontology simply means 'the study of ancient life'. Although not dealt with in detail in this book, palaeontology provides another interesting and rewarding
hobby for the amateur geologist. Fossils are a fascinating study in themselves, but they also provide an
important link with another area of geology – *stratigraphy*. During the earth's history, plant and animal
life has progressively changed and developed, so that
a particular period of time can be associated with a
particular group of fossils. This means that groups of
plant and animal fossils can be used to determine the
relative age of the sedimentary rock layers in which
they are found. By using the evidence derived from
the fossil content of sedimentary rock layers together
with that deduced from important upheavals in the
earth's crust (*structural geology*), scientists have been
able to divide geological time into four major sub-
divisions. These sub-divisions are known as **eras**, and
are the *Cryptozoic*, the *Palaeozoic*, the *Mesozoic* and
the *Cainozoic*. Each era can be further sub-divided into
periods and *epochs*. The geological time scale on page 8
relates plant and animal development to times of major
structural change in the earth's crust – periods of
major mountain-building.

Stratigraphy not only involves the history of life on
earth. It also provides evidence of the climatic and
physical conditions prevailing in each area where rocks
were being formed. At any given point in the history
of man, a number of different events take place in
different places, different societies with different
characteristics exist, and different conditions prevail
in different areas of the earth. The study of the various
branches of science, of history, geography, sociology,
meteorology and a host of other subjects, enables us
to build up a composite and overall picture of the
earth, its life and conditions, at a particular point in
history. Such a picture will be different from that of
say, a thousand years earlier, or a thousand years later.
In much the same way, the science of stratigraphy helps
to build up a composite picture of the earth, its life and
conditions, at a particular point in prehistory. The map
on page 13 is a reconstruction of what the earth
probably once looked like. It shows us that the shapes
and the positions of the continents were rather different than they are today. Similar maps can be con-

structed for any of the geological periods, using the
evidence gained from a study of the various aspects
of geology. Such reconstructions of the earth's appearance and history are paleogeographical maps.

In relatively recent times, two new sciences have
had a considerable impact on the study of geology.
Geochemistry is a highly technical aspect of the
science, and uses the most modern techniques of
analysis to study the composition of the materials
which construct the earth and moon. The science of
geophysics investigates the physical phemomena which

ABOVE **Silurian seascape**
Shallow seas and marine
conditions covered most of
the continents about
425,000,000 years ago.
They were divided into
four land masses: North
America, Europe, Asia and
the southern continents,
separated by deep oceanic
areas. This reconstruction
shows Britain under a
relatively shallow sea; the
coral reefs were probably
similar to the shallow-sea
reefs off the Australian
coast today, since Britain
is thought to have been
positioned about 30° south
of the equator at this time.

are relevant to geology. Seismology, just one branch of geophysics, is the study of earthquakes. This particular topic is instrumental in unravelling the secrets of the compositions and internal construction of the earth. Yet another branch of geophysics uses knowledge of radio-activity to determine most accurately the times at which events occurred during the history of the earth. Geophysics has also proved useful in more practical fields – in the search for oil-bearing rock formations, for example. The search for ore minerals, fuels, and the raw materials so necessary for industrial development form a part of the study known as *applied and economic geology*. The engineer, too, needs to apply geological knowledge as he designs and builds bridges, tunnels, dams, roads and runways.

The inter-relationships between these different aspects of geology have helped to provide a history of the earth from its earliest beginnings to the present day. Scientists believe that the earth probably began as a huge cloud of dust, and gradually became a smaller compacted mass. As it compacted it underwent excessive heating. The matter in the compacting mass slowly became separated into three layers, the same three layers which now compose its internal structure, and which are shown diagrammatically on page 9.

Compare the internal structure of the earth with that of a peach. When a peach is cut in half, three layers can be seen – the skin, the flesh and the stone. If the earth could be cut in half in a similar way, three distinct layers could also be seen. The thin solid crust which forms the earth's surface can be compared with the skin of the peach. Beneath this thin solid crust lies the region known as the mantle. This can be compared with the flesh of the peach. The mantle is generally solid and is kept in this state by the tremendous pressures to which it is subject. In areas where the pressure is reduced the mantle can resume its liquid state and becomes the molten rock we know as magma. The core of the earth, some 2000 miles below the surface, can be compared with the peach stone. The core is the 'home' of the earth's magnetism. Some scientists consider that the core is molten, but recent evidence suggests that it may well be solid and composed mainly of nickel-iron.

It is thought that it took over 2000 million years for the earth to assume its three-layered form, the process only reaching completion some 4600 million million years ago. The first crust to form was composed of igneous rocks. Weathering processes brought about the disintegration and erosion of these early rocks, and provided the material for the construction of sedimentary rocks. Over eons of time, life began to develop on earth. Continents and oceans formed, mountain chains were built and broken down to provide new sources of raw material for new rocks and land areas.

Under certain specialized conditions it is known that oil and gas deposits were formed. Deposits of gold, silver, lead and many other minerals were formed and deposited in various localities. Slowly, the earth as we know it was forged from the raw materials of the initial clouds of dust. Today we depend for our existence on the minerals and rocks that were formed long before man set foot on the earth.

What evidence that we have of the internal structure of the earth is largely derived from the

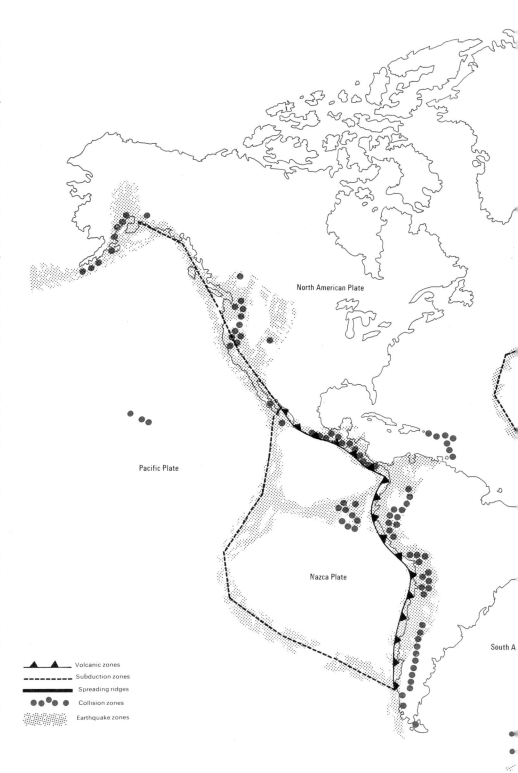

North American Plate

Pacific Plate

Nazca Plate

South A

▲▲ Volcanic zones
---- Subduction zones
—— Spreading ridges
●●●● ● Collision zones
░░░ Earthquake zones

study of measurements taken during earth movements.

There are three main types of wave generated from an earthquake centre – 'P' waves, 'S' waves and 'L' waves. 'P' waves are compressional waves, 'S' waves are shear waves. Both travel through the earth. 'L' waves, on the other hand, travel round the outer layer of the earth, and it is these waves which are the destructive agents in any great earthquake.

It is from the study of 'P' waves and 'S' waves particularly that information concerning the interior of the earth has been obtained. Experiment has shown that when 'P' waves pass through an elastic solid their

ABOVE **Tectonic plates** The crust of the earth consists of a number of huge plates, one adjacent to the next and 'floating'. These plates are in constant motion. The areas where the plates meet form lines of weakness in the earth's crust, with earthquake and volcanic activity. The theory of tectonic movement is still not fully developed.

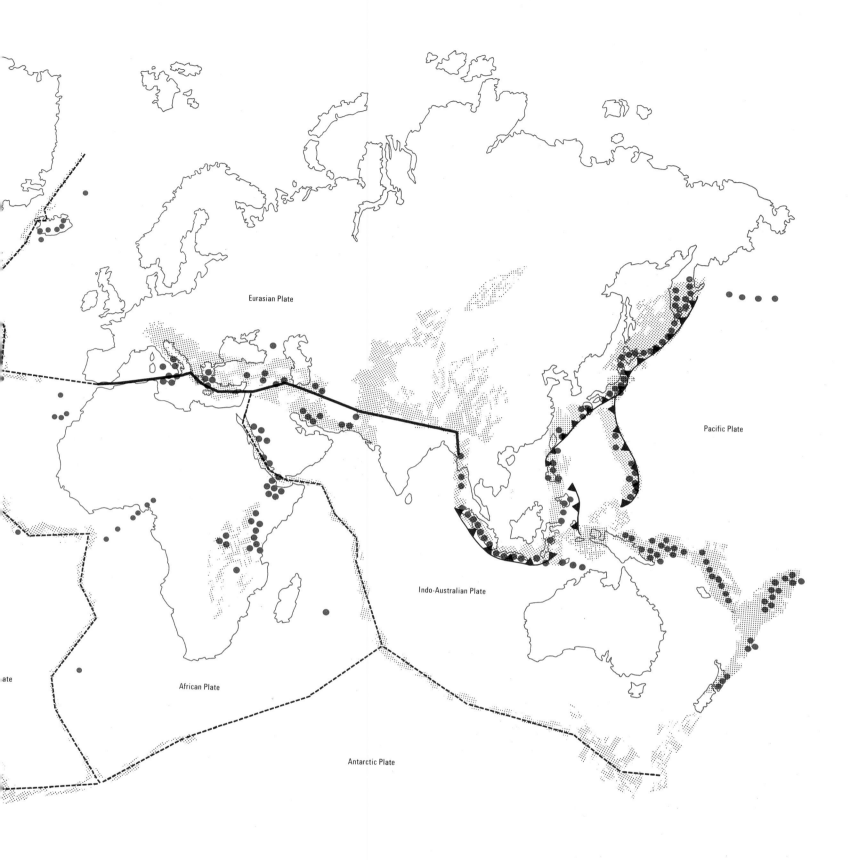

Eurasian Plate

Pacific Plate

Indo-Australian Plate

African Plate

Antarctic Plate

rate of travel varies directly with the resistance the solid offers to compressive forces and shear, and inversely as the density of the medium varies. 'S' waves behave in a similar way. Their velocity varies directly with the resistance to shear and indirectly in relation to density. It follows that in rocks of the same density the velocity of the waves will increase with the rigidity and incompressibility of the rock.

The Austrian scientist, Mohorovičić, recognized that the variations in velocity shown by earthquake waves were a consequence of the layered character of the crust. He showed that the behaviour of the waves

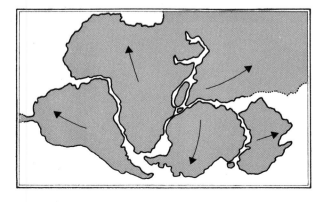

LEFT **A reconstruction of the continents several millions of years ago** Some geologists think that the continents were once joined together in one huge land mass, and that they slowly drifted apart. The lines along which they separated were lines of tectonic weakness in the earth's crust. The arrows show the drift direction.

indicated a two-fold division of the crust into an upper layer, which seemed to have the properties of a granite, and a lower layer, which seemed to have the properties of a basalt. Since the upper layer was rich in silica and alumina, it was given the name 'sial'; the lower layer, rich in silica and magnesia, was given the name 'sima'. The continental land masses appear to be mainly composed of sial; the ocean floor and the oceanic islands appear to be mainly composed of sima.

At a depth of approximately 33 kilometres, there is a marked change both in the paths of the 'P' and 'S' waves and in their velocities. This suggested that there is a marked change in rock type at that depth. The boundary has been named the Mohorovičić Discontinuity, and marks the limit of the earth's crust.

Below the crust the rock type appears to have the properties of an ultra-basic igneous rock. The rock type seems to vary with depth. The evidence on this is that the density varies from 3.32 just below the crust to 5.66 at the deepest extent. The velocity of the 'P' waves varies considerably also, from 8 kms s^{-1} to 13.64 kms s^{-1}. This zone seems to extend to a depth of 2900 km and is known as the mantle.

Below the mantle, the behaviour of the 'P' and 'S' waves again varies, and this is also ascribed to a change in the rock type. The surface at which the change occurs is known as the Oldham-Gutenburg Discontinuity. Below this discontinuity is the outer core, which has a density of 9.71. Between the outer core and the inner core lies the zone of transition. The inner core has a density of 16.

ABOVE **Ayer's Rock, Northern Territory, Australia** An example of sedimentary rock formed from glacial deposit, Ayer's Rock is a monolith – a large rock above a very flat surrounding surface, which has been eroded away.

LEFT **Rock formations, near Kalbaari, West Australia** These rocks show clearly how wind and water have eroded the sandstone, leaving a prominent overhang.

LEFT **Sandstone boulders, National Park, Rhodesia** There is some dispute among geologists as to how this formation occurred, but it has been suggested that these boulders are the result of erosion. If such boulders are preserved in the geological strata, they form the sedimentary rock known as a conglomerate.

LEFT **Earth pyramids
near Bolzana, North Italy**
An unusual sculptural effect;
most of the rock has been
washed away by rain water
leaving the pillars which
have protective boulders as
caps.

ABOVE **Natural arches, Arizona National Park** The extent to which wind and water can carve out exposed rock is clearly shown by this rock arch in sandstone.

The behaviour of the 'P' and 'S' waves as they pass from the mantle to the core suggested that there was a change in the state of the rocks below the discontinuity. Opinion was that the core was in a liquid state. However, other evidence has been found in recent years, and the Danish seismologist, Inge Lehmann, put forward a theory that there was a further discontinuity at the inner core. Seismic measurements gathered from the explosion of nuclear devices seems to confirm this theory. Many scientists feel that the inner core is a solid body, and that its general composition is that of a nickel-iron compound.

The growth of geology

Geology is one of the youngest of the sciences; in fact, it was not considered a science at all until the end of the eighteenth century. The word geology was coined by a French scientist and comes from *geo*, which means earth, and *logos*, meaning discussion or discourse. However, 'geologists' have endeavoured to explain natural phenomena from the earliest times. The early Greek philosophers, Thales, the Pythagoreans, Plato, Aristotle and Herodotus all developed theories about the earth, its shape and the ever-changing nature of its surface. Our earliest ancestors used the rocks and minerals of the earth as ornaments, weapons and fuels. Most of history's great scientists have considered at least one facet of geology.

This introductory chapter has set out to describe briefly the science of geology and its part in the sum knowledge of our planet and its riches. The following chapters describe in greater detail the basic rock types, their structures and formation processes; the minerals and rocks which are so important to our existence; and the rare and precious materials we know as gemstones. It is not a scientific book, in the sense that it contains erudite discussions, complex explanations and complicated formulae; but it does try to describe the wealth of interesting knowledge and information that lies, literally, beneath our feet.

RIGHT **The birth of a new island** In November 1963 the crew of a small vessel off South Iceland suddenly saw a great column of smoke and ash emerge from 100 fathoms of the Atlantic and rise to a height of several thousand feet. This volcanic eruption resulted in the formation of a new island, which was named Surtsey. The volcano is still active and the island is still growing (see page 39).

Chapter 1
MINERALS

The rocks which make up the earth's crust and inner layers vary so greatly in appearance that it seems unreasonable to characterize them all under one general term. But each of the rocks shares a common feature – it is composed of one or more of the basic substances we call minerals.

While the minerals which constitute the rocks are not always present in the same proportions, minerals themselves are essentially uniform in composition. They are always naturally occurring substances. They have an internal atomic structure, which is often expressed in outward crystal form – cubic halite crystals and hexagonal quartz crystals are examples. Minerals are generally chemical compounds consisting of two or more elements in combination. Quartz, with a combination of silicon and oxygen atoms; and salt (the mineral halite) with the atoms of sodium and chlorine, are relatively simple examples of such minerals. Other minerals are extremely complex in structure. The mineral mica, for example, is a compound of potassium, magnesium, iron, aluminium, silicon, oxygen, hydrogen, and fluorine.

There are more than one hundred elements known to scientists; some of them have been known for thousands of years, others have only been discovered in highly complex and technical nuclear laboratories. Elements are substances that cannot be separated into simpler forms of matter by ordinary means. Each element is given a shorthand symbol to represent it. For example, Pb is the symbol for lead, Cu is the symbol for copper and Sn is the symbol for tin. These rather odd symbols are derived from the Latin names for the substances. In fact the Romans knew nine of the elements – lead (plumbum), copper (cuprum), tin (stannum), carbon (carbo), sulphur (sulfur), gold (aurum), silver (argentum), iron (ferrum) and mercury (hydrargyrum). Hence the symbols Pb, Cu, Sn, C (carbon), S (sulphur), Au (gold), Ag (silver), Fe (iron) and Hg (mercury). Other elements are represented by symbols denoting their modern names. O (oxygen), Pt (platinum), Si (silicon) and Al (aluminium) are examples.

The shorthand symbols can be combined to represent combinations of elements; for example, SiO_2 is the shorthand symbol for an oxide of silicon, better known as the mineral quartz; and NaCl is the symbol for sodium chloride, better known as salt, or the mineral halite. The complex structure of minerals such as mica can be shown by using the individual symbols together – $K(Mg,Fe)_3(AlSi_3)O_{10}(OH,F)_2$.

To return for a moment to elements. Of the hundred

A group of amethyst crystals Amethyst is a purple variety of the mineral quartz, and owes its colour to the presence of manganese impurities. These crystals show the typical hexagonal prism of the quartz group of minerals. Amethyst crystals are often cut and polished for use as gemstones.

or so known elements, only about ninety are found naturally in the earth's crust. Of these ninety, eight are vastly more abundant than the others, adding up to almost 99% of the whole. The eight predominant elements are oxygen (O, 46.59%), silicon (Si, 27.72%); aluminium (Al, 8.13%); iron (Fe, 5.01%); calcium (Ca, 3.63%); sodium (Na, 2.85%); potassium (K, 2.60%); and magnesium (Mg, 2.09%). These figures show that three-quarters of the earth's crust consists of oxygen and silicon. Most of the earth's minerals are compounds made up of these elements in combination with one or more of the other six, sometimes with small amounts of the other, rarer, elements.

It was stated earlier that an element is defined as a substance that cannot be separated into simpler forms of matter by ordinary means. Obviously, we could take a sample of an element, say a bar of iron, and physically divide it into two equal samples. We could then repeat the process over and over again – assuming we had the technical apparatus necessary to divide minute quantities. Clearly, we could carry on the division process until we had a quantity which, however hard we tried, we could not divide further. We would have the smallest fraction of the element possible. In fact, it *is* possible to divide this smallest fraction – but if we did this, we would no longer have a

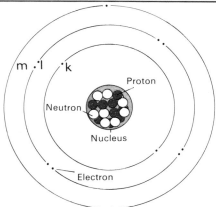

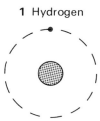
1 Hydrogen

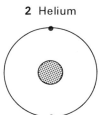
2 Helium

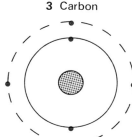
3 Carbon

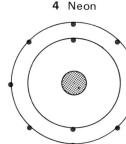
4 Neon

ABOVE CENTRE **Diagrammatic representaion of a number of different atoms 1** *Hydrogen* The hydrogen atom has only one electron in the k shell. **2** *Helium* has two electrons in the k shell. **3** *Carbon* has two electrons in the k shell and four in the l shell. **4** *Neon* has two electrons in the k shell and eight in the l shell. **5** *Silicon* has two electrons in the k shell, eight in the l shell and four in the m shell. **6** *Copper* has two electrons in the k shell, eight in the l shell, eighteen in the m shell and one in the n shell.

RIGHT **Crystals of tourmaline** These crystals show the three-sided columnar structure which is an important characteristic of tourmaline. Tourmaline occurs in a variety of colours. Some specimens are colourless, many are in shades of blue and black.

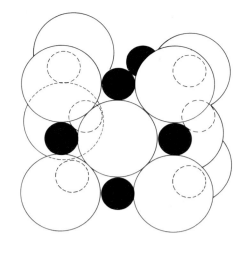

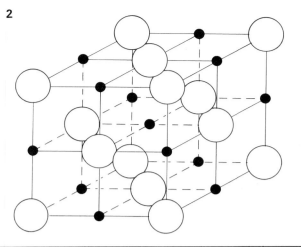

LEFT **The atomic structure of sodium chloride 1** Close packing of ions shows the sodium ions (solid black circles) fitting the interspaces between the chlorine ions (open circles). **2** Eight unit cells showing the sharing of ions between adjacent cells.

BELOW **A molecule of carbon dioxide** showing the sharing of electrons between ions.

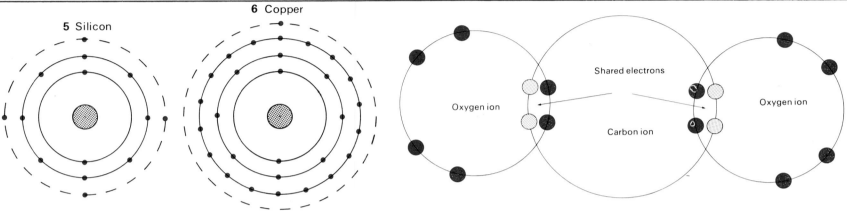

sample of the element. In fact, we would have split an atom of the element, and this changes its characteristics and properties. The smallest fraction of an element which can exist by itself and still retain the element's chemical properties is the *atom*. Atoms of elements can combine together under certain circumstances to form tiny fractions of new substances – compounds; the smallest fraction of a substance consisting of two or more elements chemically combined together which can exist without altering the substances composition or chemical properties, is called a *molecule*. The chemical shorthand discussed earlier refers to the number of atoms required to form a given material. For example, the mineral halite – salt – has the chemical formula NaCl. This tells us that one atom of sodium (Na) combines with one atom of chlorine (Cl) to form one molecule of sodium chloride (NaCl).

Another concept which is very important to mineralogists is atomic structure. This describes how the individual atoms are actually constructed and how atoms link together. Some minerals are composed of identical atoms but the atoms link together differently, thus forming minerals with very different characteristics. For example, diamond and graphite are both made up of atoms of carbon, yet diamond is whitish and extremely hard (the hardest naturally occurring substance known to science), while graphite is black and very soft (so soft that it is used as an excellent industrial lubricant).

The structure of matter

Many people believe the concept of atoms and atomic structure is a relatively modern one, yet as long ago as 450 BC Greek philosophers conceived the idea that all matter was composed of minute units which they called atoms. This idea persisted in the minds of philosophers, yet a precise theory concerning the constitution of matter was not formulated until 1802 when Dalton propounded his Atomic Theory. This theory claimed that all matter was made up of small particles called atoms which were indestructible and which could not be created; that the atoms of a particular element were all alike but differed from the atoms of all the other elements; and that chemical combination took place between small whole numbers of atoms. Dalton's theory represented a great step forward in scientific thinking and laid the foundation stones for modern concepts of atomic structure. However, progress was slow, and it was not until

BELOW **Sulphur** A rather shapeless encrusting mass of this native mineral which may be recognized by its distinctive yellow colour and its light weight.

1895 that ideas about the internal structure of the atom began to appear. In 1913, Neils Bohr, a Danish physicist, put forward his idea that the atom was composed of three distinct units – protons, neutrons and electrons. He suggested that these units were related to each other by using a demonstration model which looked rather like a small solar system.

The nucleus of this system is composed of protons and neutrons, which have an equal mass. Revolving around the nucleus, in orbit, are the electrons. They are much smaller than the protons and neutrons, having a mass about 1/1860th of them. Each proton in the nucleus carries a tiny positive electrical charge. Each electron revolving around the nucleus carries a tiny negative electrical charge. Because there are as many electrons as there are protons, the charges 'cancel each other out', so that overall, the atom is electrically neutral. These electrical charges are extremely important in the formation of chemical compounds, as will be discussed later.

The orbits in which the electrons travel around the central nucleus are generally known as electron shells, and given the symbols k,l,m,n and so on. Each shell can carry a specific maximum number of electrons – but each shell does not necessarily carry its full 'quota'. The three inner shells – the k,l and m shells, appear to be the most important and can carry a maximum of 2, 8 and 18 electrons. These electrons orbit with tremendous speed. If it were possible to see an atom, the electrons would appear to form a solid shell, so fast do they travel. An atom is 'held

LEFT **Chalcopyrite** Copper pyrites occurring as cube- and wedge-shaped brass-yellow crystals. These minerals are about 35% copper.

RIGHT **Malachite** Malachite is yet another important source of copper since it contains some 60% copper ore. Malachite is usually this particular shade of green and often occurs shaped like a bunch of grapes.

FAR RIGHT **Azurite** Azurite is another important commercial source of copper, and is often found associated with malachite. Although azurite generally occurs in the massive state, it can occur as long, blue, prism-shaped crystals.

BELOW LEFT **Galena** The cubic crystals, which are lead grey in colour, are typical of the mineral galena, which is an important source of lead. Galena is often known as 'blue lead'. The white crystal is calcite.

together' by tremendous forces. Some idea of these forces can be grasped when one realizes that a nuclear explosion results when atoms are 'split' and the forces released.

The formation of compounds

For atoms to join together to form compound units, some attraction must develop between them. In fact, the force is an electrical one and is the result of an atom gaining or losing electrons from its electron shells. The addition or loss of electrons destroys the electrical balance within the atom, for the positive and negative electrical charges are balanced because the numbers of protons and electrons are identical. Thus the addition or loss of electrons causes the atom to aquire a negative or positive charge. For example, the mineral halite, which we have already mentioned, consists of the two elements sodium and chlorine. Sodium atoms have the two inner electron shells complete, but only carry a single electron in the outer m shell (the electron configuration is thus 2.8.1). On the other hand, chlorine atoms have seven electrons in the m shell (the electron configuration is 2.8.7). An element with a more nearly complete outer shell will attract electrons from an element with a less complete outer shell. Thus, the chlorine atom will attract the single electron from the sodium atom. This transfer results in sodium, which has lost an electron, having a positive electrical charge; while chlorine, which has gained an electron, becomes negatively charged. The two atoms are now called *ions* because they carry an electrical charge, and unite to form a molecule of sodium chloride – halite.

Ionic bonding One of the laws of electricity states

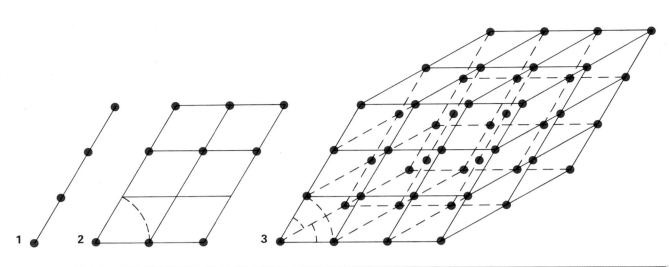

RIGHT **One, two and three dimensional patterns 1** A linear pattern. **2** A two dimensional repeat pattern. **3** A three dimensional space pattern.

FAR RIGHT **Three space lattices 4** Orthorhombic. **5** Monoclinic. **6** Hexagonal. The solid circles represent atoms, molecules or partial molecules distributed in space to form three dimensional lattices. There are fourteen such forms which embody all ways of arranging points in space.

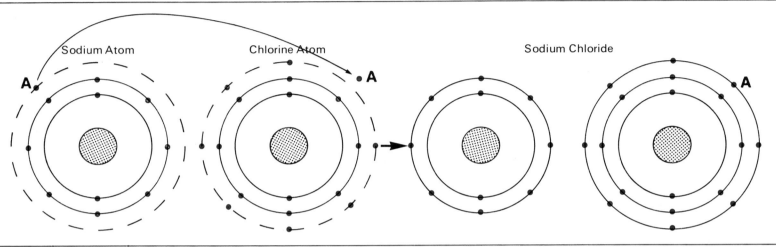

Sodium Atom   Chlorine Atom   Sodium Chloride

ABOVE **Ionic bonding in sodium chloride** The electron A in the outer shell of the sodium atom transfers to the outer shell of the chlorine atom, the atoms are ionized and join to form the sodium chloride.

RIGHT **Pyrites** Iron pyrites occurs in cube-shaped crystals.

that 'like charges repel, unlike charges attract'. Thus there is a force of attraction between the oppositely charged ions, and the atoms bond together. This type of bonding is known as ionic bonding and results in each ion being surrounded by ions of opposite charge. The potential to transfer electrons is known as *electrovalence* and produces closely packed masses of ions described as ionic agglomerates. Most of the natural minerals have this kind of atomic bonding. **Homopolar bonding** results from the sharing of one or more electrons by adjacent atoms. For example, an atom of fluorine has seven electrons in its outer shell. Six of these electrons are paired, leaving a single electron. When two fluorine atoms approach each other, their outer shells penetrate each other and the two single electrons pair off. This electron pair is shared by the two atoms, which are now joined to form a molecule of fluorite.

Diamond is another mineral which shows homopolar bonding. A carbon atom has four electrons in its outer shell and can form four homopolar bonds with four other carbon atoms thus producing a tetrahedral structure. This is an extremely strong structure and accounts for the hardness of diamond. Graphite, on the other hand, consists of carbon atoms lying in planes, each atom lying at the corner of a regular hexagon. The result of this is that graphite is very soft, and the layers slide easily over each other (which accounts for graphite being a good lubricant). **Metallic bonding** is a bonding characteristic of metals. Closely packed positively charged ions are enclosed by a cloud of negatively charged free electrons. Cohesion is achieved through the electrical

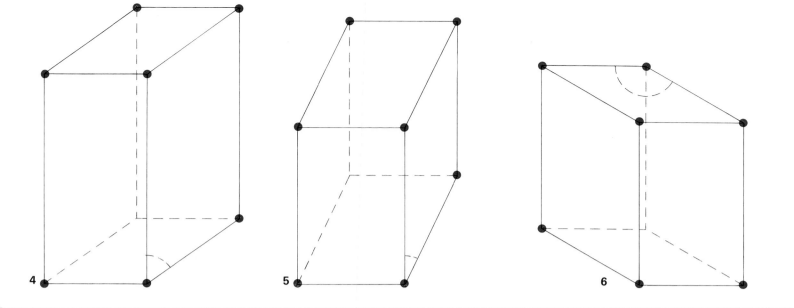

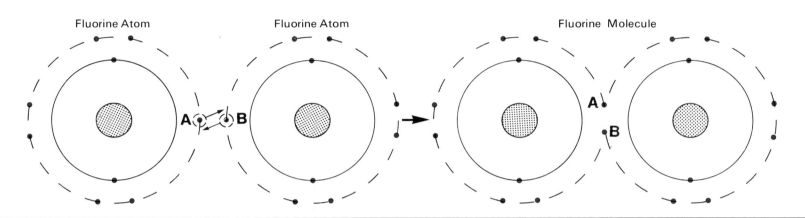

Fluorine Atom   Fluorine Atom   Fluorine Molecule

attraction between the two oppositely charged groups of electrons. The electrons in the surrounding cloud can easily be agitated, and it is this property that makes metals such good conductors of electricity.

The atoms forming a compound are held together by these various forms of bonding. They are held in a distinct geometric pattern. Substances like this can continue to crystallize, still maintaining the geometric shape. Under ideal conditions, such substances form crystals.

Crystallography

The important feature of a crystalline substance is the regularity of the arrangement of the ions or molecules from which it is built. This distribution of building 'units is in the form of a three-dimensional symmetrical' pattern in space. All crystals can be placed in one of seven crystal systems, each characterized by a specific group of symmetry elements. The three elements of symmetry are the plane of symmetry, the centre of symmetry and the axis of symmetry. A plane of symmetry divides a crystal into two halves, each of which is the mirror image of the other. A centre of symmetry is present when a line passed from one face to another on the opposite side of the centre emerges at a point similarly disposed to the one at which it entered. A consequence of the centre of symmetry is that there are pairs of parallel faces present in the crystal. An axis of symmetry is a direction about which a crystal can be rotated so as to bring a similar face into view more than once in a revolution of 360°.

Such axes may be digonal, when a similar face is seen twice in a complete revolution of 360°; trigonal,

ABOVE **Homopolar or covalent bonding in fluorite** Each atom of fluorine has seven electrons in its outer shell. Two atoms approach each other so that the single electrons can join to form an electron pair as in A and B.

The Elements of Symmetry of the Holohedral Class of each of the seven Crystal Systems

| SYSTEM | Planes | Centre | Axes | | | | Examples |
			Digonal	Trigonal	Tetragonal	Hexagonal	
Cubic	9	C	6	4	3	—	Galena, Garnet, Fluorite, Rocksalt, Diamond
Tetragonal	5	C	4	—	1	—	Zircon, Idocrase, Wulfenite, Chalcopyrite
Orthorhombic	3	C	3	—	—	—	Topaz, Barytes, Olivine
Hexagonal	7	C	6	—	1	—	Beryl, Emerald
Trigonal	3	C	—	1	—	—	Calcite, Tourmaline
Monoclinic	1	C	1	—	—	—	Orthoclase, Crocite, Azurite, Gypsum
Triclinic	0	C	—	—	—	—	Amazonite, Kyanite

LEFT **Elements of symmetry** Within each of the seven crystal systems there are classes with higher degrees of symmetry than the minimum. The class with the highest symmetry in each class is the holohedral class.

LEFT **Antimonite (Stibnite)**
The usual crystal form of
the mineral antimonite is
long, striated columns.
Antimonite is the chief
source of the metal anti-
mony, used mainly in the
production of alloys and in
the ceramic industry.

BELOW **Pyrites on calcite**
The brass-yellow colour of
the pyrite crystals is
typical. Here the crystals
are in a finely disseminated
form. The streak of pyrites
is greenish black and it
has a hardness of $6\frac{1}{2}$. The
calcite beneath is white.

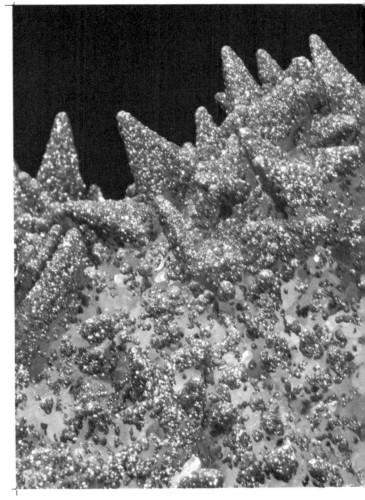

when a similar face is seen three times in a complete
revolution; tetragonal, when a similar face is seen four
times; and hexagonal, when a similar face is seen six
times.

In each of the seven crystal systems there is a certain
minimum amount of symmetry required. But within
each system there are classes with higher degrees of
symmetry than the minimum. The class which shows
the highest degree of symmetry within each system is
known as the *holohedral* class.

In nature, crystals do not as a rule crystallize in
good geometric form. Due to a variety of adverse
conditions during the crystallization processes the
same form of face in the same crystal may show
marked differences in development. However, well-
formed crystals do sometimes occur in nature, as is
apparent from the photographs in this chapter.

LEFT **Fluorite (Fluorspar)**
The cubic form of fluorite
is clearly shown. Although
fluorite is found in a variety of colours, purple is the
most common. However,
the streak of fluorite is
always white, like the pure
form of the mineral.

Each one of the 2000 known minerals has its own
distinctive set of properties and characteristics by
which it can be distinguished from the others. The
characteristics are determined by both the atomic
composition and the atomic pattern which the mineral
possesses – we have seen, for example, how the
atomic bonding of carbon atoms is responsible for the
different properties of diamond and graphite.

The characteristics by which minerals can be
recognized fall into two categories; identification by
visual observation or by means of simple tests; and
identification by the application of specialized knowledge or by means of complex scientific tests using
sophisticated equipment. Luckily, most of the common
minerals can be identified by tests in the first category.

The simplest tests can be further sub-classified into
four groups:

ABOVE **Tourmaline with quartz** Tourmaline occurs in three-sided crystals (see page 20) often striated along the length and terminated by three polygonal faces. Here it occurs with the mineral quartz. Tourmaline has a wide colour spectrum and can vary quite substantially in colour within a single crystal.

1. tests for characteristics depending on light: colour, streak, lustre.
2. tests for characteristics depending on the way the atoms are linked together: habit, hardness, fracture, cleavage.
3. tests for characteristics depending on atomic weight and packing; specific gravity.
4. tests for characteristics depending on the senses: taste, odour, feel.

Tests based on electrical and magnetic properties are useful in a limited number of minerals.

Colour Some minerals have a single characteristic colour – the green of malachite, the blue of azurite, the yellow of pyrite, the silver-grey of galena and the dull black of graphite are examples. Most minerals, however, show a variable range of colours. This is due to small amounts of impurities which are present in the mineral. Some minerals – opal, for example – are multi-coloured; others appear as differently coloured forms of the same mineral. For example, ruby and sapphire are coloured forms of the mineral corundum. Exposure to air may change the original colour of a mineral – siderite is grey-brown when first mined, but darkens on exposure to air; bornite and chalcopyrite tarnish to iridescent purple colouring. Colour, therefore, is not the most satisfactory criterion to use in mineral identification, for there many similarly coloured minerals, and colour variations are frequent within a single mineral type.

Streak is the powder mark left by a mineral as it is drawn across an unglazed piece of porcelain – a streak plate. Streak has several uses as a diagnostic test and it is important that the streak test is used in amateur mineral identification. The impurities which cause subtle colour changes in some minerals are present in only small quantities and so they are not represented in a single powder mark. In some minerals, the streak test is diagnostic; for example, the black-brown mass of kidney iron ore (haematite) always gives a cherry-red streak; the scarlet-coloured cinnabar gives a scarlet streak.

Lustre is the way in which light is reflected from the surface of a mineral. Some minerals have the appearance of metal – galena, magnetite, pyrite and graphite are examples. This appearance is known as metallic lustre. However, the majority of minerals have non-metallic lustre. They can be further sub-classified into a number of groups of which the following are important: brilliant lustre, vitreous lustre, resinous lustre, pearly lustre and silky lustre. Minerals with brilliant lustre include zircon, and of course diamond. Vitreous lustre is the most common non-metallic lustre and minerals in this group include quartz, garnet, tourmaline. Minerals with resinous lustre have a wax-like appearance; examples include realgar, opal, amber, sphalerite and sulphur. Selenite, gypsum and talc have the rainbow effect which is the characteristic of minerals having pearly lustre; while the shimmering softness of asbestos and satin-spar gypsum is one of the characteristics of silky lustre.

Light The extent to which a mineral permits the passage of light is another simple test. Opaque minerals do not allow any light to pass through, translucent minerals allow some light to pass through, transparent

LEFT **Beryl crystals** The green variety of beryl is better known as the gemstone emerald, the blue variety is known as aquamarine. Pink beryl is known as morganite, and brown beryl is known as heliodor. Beryl is generally found in the form of six-sided crystals.

ABOVE **Garnets in a mica schist** Crystals of garnet are many-sided and red-brown in colour. They are often found embedded in metamorphic rocks, such as the mica schist illustrated here. The smaller dark-coloured patches among the white mica are also garnets.

TOP **Kyanite (Disthene)** Kyanite may occur as long bladed crystals, which are blue in colour. It is the particular colour of this mineral that is very characteristic, for kyanite very rarely shows crystal form. It is found in association with metamorphic rocks.

minerals allow almost all light to pass through. Some transparent minerals – calcite and zircon, for example – demonstrate the phenomenon of double refraction; if an object is observed through the mineral, a double image is seen. Some minerals have the ability to change colour under certain lighting conditions. This property is known as pleochroism (see page 93).

Habit is the growth form of a mineral assumed as it grows in its particular environment. Some minerals occur in masses, others occur as individual minerals in a particular form – as crystals. The various crystal systems are shown on page 26. Crystal shapes are often valuable clues in the identification of minerals.

Hardness The hardness of a mineral largely depends on the strength of the atomic bonding forces. The harder minerals have atoms which are very tightly bonded together. Hardness is measured as the resistance to scratching by various implements (kits of hardness pencils are available to the amateur geologist).

The hardness of minerals is measured on a scale devised by the German mineralogist, Friedrich Mohs, and known as Mohs' scale of hardness. This scale grades hardness from very soft, grade 1, to very hard, grade 10.

The hardness test is not easy to apply, but once mastered it proves a useful aid to mineral identification. Slight variations in hardness are to be expected since there is often some degree of variation in the atomic bonding in a mineral. Generally, if the mineral under investigation will scratch glass then it has a hardness value of 6 or more. If the mineral can be scratched by a knife (generally only tiny scratches in unobtrusive parts of the mineral are made) but not by a finger nail, then it has a hardness value of between 2 and 6. If the mineral can be scratched and powdered by a coin, the hardness is between 2 and 4. If, however, the *coin itself* is scratched, the hardness is between 4 and 6.

Tenacity Many minerals will shatter into fragments

MOHS' SCALE OF HARDNESS

Hardness Scale	Representative Mineral	Other Minerals	Quick Test
10	Diamond	—	Will Scratch Glass
9	Corundum	—	
8	Topaz	Beryl, Spinel, Zircon	
7	Quartz	Garnet, Staurolite, Tourmaline	
6	Feldspar	Turquoise, Rutile, Celsian	Can be scratched with steel knife
5	Apatite	Bornite, Smithsonite	
4	Fluorite	Chalcopyrite, Malachite, Azurite	Can be powdered by scratching with coin
3	Calcite	Argonite, Barytes	
2	Gypsum	Salt, Sylrite	Can be scratched with finger nail
1	Talc	Aluminite, Limonite, Ulexite	

when tapped lightly with a geologist's hammer. This property is known as brittleness. A few minerals – gold, silver and copper, for example – can be flattened out into thin sheets, without shattering or fracturing. This property is known as malleability. Ductility, the property which allows a mineral to be drawn out into a long, thin wire, is a property connected to malleability. Small pieces of certain minerals can be 'bent'. Mica is an example of a mineral which returns to its original form after being bent, and is therefore said to be elastic and flexible. Talc, asbestos, gypsum and chlorite are all flexible minerals, as they can be bent, but they are non-elastic, since they remain in their new positions, even when the bending pressure is removed.

Fracture Fracture is an irregular break in a mineral. Both quartz and obsidian show a fracture pattern rather like ripples on the surface of a pond when it is disturbed by a stone. This 'ripple' fracture is known as conchoidal fracture. Copper and silver have a

LEFT **Feldspar** This picture shows crystals of white plagioclase feldspar, one of the most common of all the rock-forming minerals. These crystals show the two cleavages at right angles, which is one of the diagnostic features to be looked for in this mineral.

BELOW **Pyromorphite** This mass of prism-like crystals is an example of one of the lead minerals – hence its more common name, yellow-green lead ore. Its colour and its high specific gravity are important aids to its identification.

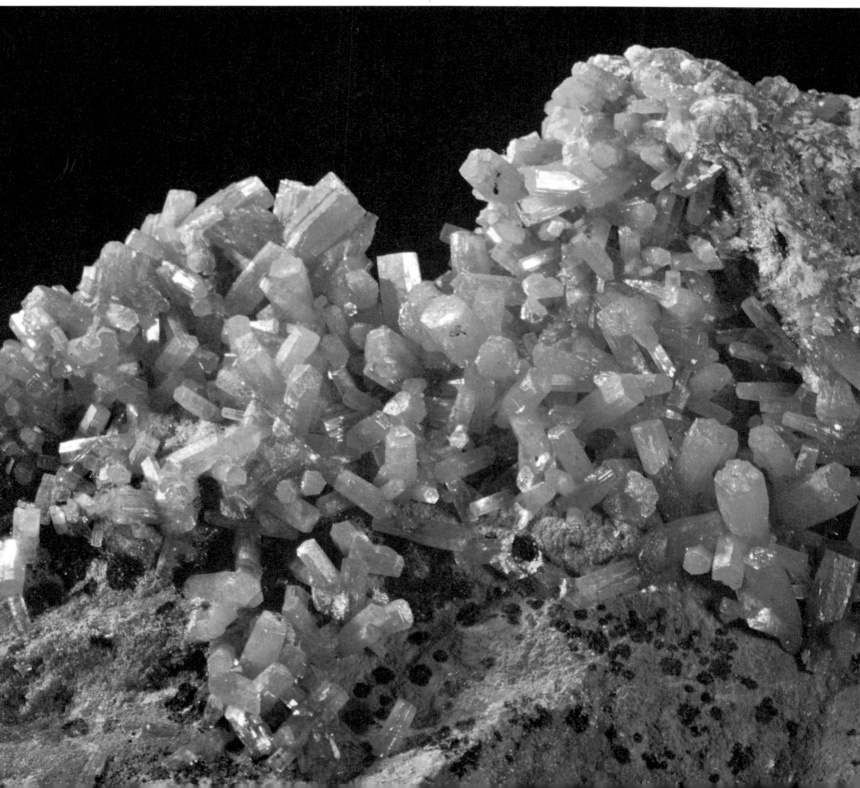

RIGHT **Halite (Rock salt)**
Halite can occur in massive form, as shown here, or it can occur in the form of cubic crystals. Halite is usually translucent, and this, together with its distinctive taste, is diagnostic.

BELOW **Cleavage in different minerals** Cleavage is a property determined by the geometric arrangement of atoms in a mineral. The number of cleavages and the angles between them are always characteristic of a particular mineral.

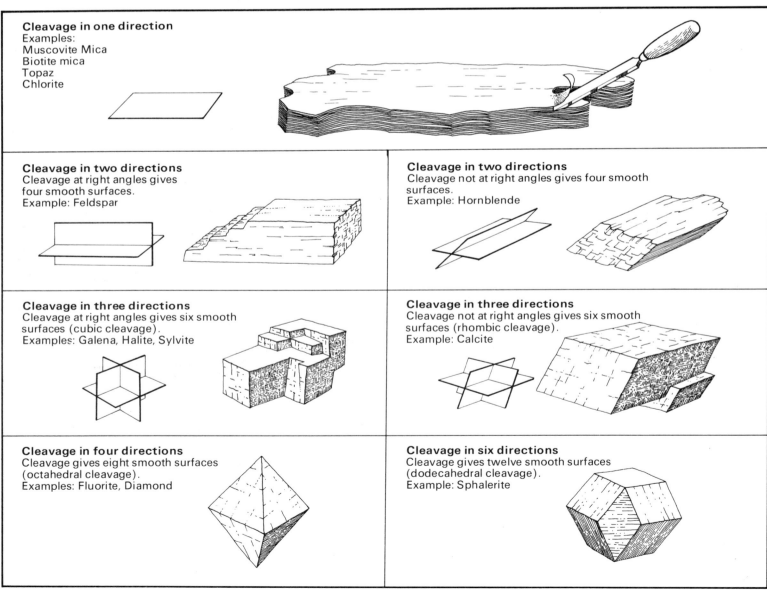

Cleavage in one direction
Examples:
Muscovite Mica
Biotite mica
Topaz
Chlorite

Cleavage in two directions
Cleavage at right angles gives four smooth surfaces.
Example: Feldspar

Cleavage in two directions
Cleavage not at right angles gives four smooth surfaces.
Example: Hornblende

Cleavage in three directions
Cleavage at right angles gives six smooth surfaces (cubic cleavage).
Examples: Galena, Halite, Sylvite

Cleavage in three directions
Cleavage not at right angles gives six smooth surfaces (rhombic cleavage).
Example: Calcite

Cleavage in four directions
Cleavage gives eight smooth surfaces (octahedral cleavage).
Examples: Fluorite, Diamond

Cleavage in six directions
Cleavage gives twelve smooth surfaces (dodecahedral cleavage).
Example: Sphalerite

different fracture pattern. When fractured they yield very sharp and jagged edges, and this fracture pattern is known as hackly fracture.

Cleavage Cleavage is a smooth break in the mineral and occurs as a consequence of regularly spaced sheets of ions with weak bonding between the sheets, but stronger bonding within the sheets. The mineral tends to split along a clearly defined plane (see page 31).

Specific gravity This is the term used to express the density of a mineral. Density is partly a result of the weight of the atoms (atomic weight) and partly a result of the way in which the atoms are packed together. The more tightly the atoms are packed, the higher will be the density, the higher will be the specific gravity. Two pieces of mineral may be of similar size, yet have different specific gravities – just as a golf ball and a table tennis ball are of similar size yet have different densities and specific gravities.

There are a number of sophisticated methods of determining specific gravity of a given sample. One of the simplest ways is to weigh the sample using spring balance, then to reweigh the sample just submerged in water. The two readings are then substituted in the formula:

$$\frac{\text{weight of mineral in air}}{\text{weight of mineral in air} - \text{weight of mineral in water}}$$

For rough and ready tests, it is often sufficient to estimate the specific gravity of a mineral in terms of 'light', 'heavy' and 'medium' by 'weighing' the mineral in the hand. Gold, silver and copper have very high specific gravities. Galena, cassiterite, haematite and pyrite, with specific gravities of $7\frac{1}{2}$, $6\frac{1}{2}$, 6 and 5 are considered to have high specific gravities. Barytes ($4\frac{1}{2}$), malachite (4) and augite ($3\frac{1}{2}$) are considered to have specific gravities in the medium range. Salt, gypsum, sulphur, graphite, calcite, quartz and feld-spar are all in the lower range of specific gravities.

Reaction to senses The set of properties which can be detected by taste, smell and touch can be an extremely useful aid in the identification of several minerals. The salty taste of halite, the bitter taste of sylvite, the clayey taste of kaolinite, the sulphurous odour of newly-broken pyrite, and the greasy, soapy feel of graphite, talc and chlorite are good examples.

Electricity and magnetism Some minerals aquire a small electrical charge when they are rubbed with silk or nylon, and this charge enables them to attract dust, pieces of paper and other small objects. Of the common minerals, only sulphur has this property, but topaz, tourmaline and diamond also have the ability to receive a small electrical charge. Of the common minerals, only magnetite will react to a magnet.

Acid tests Reaction to acids is another simple test

BELOW **Rutilated quartz** A quartz crystal containing radiating needles of the mineral rutile which have developed from the massive titanium-bearing ore at the base of the quartz. Rutile is a semi-precious gemstone.

The examination under special conditions of thin sections of rock can help a mineralogist to identify the physical properties of the specimen. The mineralogist uses a special microscope known as a petrological microscope, which uses polarized light. The wafer-thin rock specimen can be viewed under plane polarized light, or under crossed polarized light. The colour characteristics displayed by minerals when viewed under these two conditions allow identification of the specimen.

LEFT **Garnet-mica-schist View A under plane polarized light** Large, well-shaped garnets are seen in a finer ground mass of mica (muscovite), quartz and iron oxide.

BELOW LEFT **View B under crossed polarized light** The garnets are recognized by the typical black colour. Muscovite micas are the highly-coloured minerals, the quartz minerals are grey.

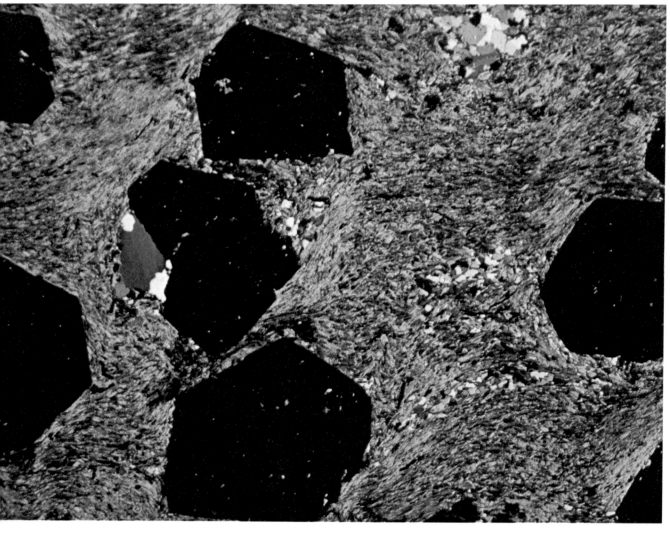

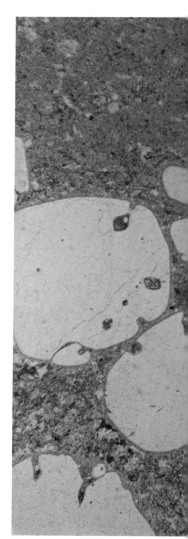

RIGHT **Olivine-basalt View A under plane polarized light** The large, cracked minerals are olivine; the smaller minerals which make up the ground mass include feldspar, pyroxene and glass.

FAR RIGHT **View B under crossed polarized light** The olivine becomes yellow in colour, or blue. The smaller pyroxene crystals are now brown, iron oxide is black.

BELOW **Quartz-porphyry View A under plane polarized light** The large quartz crystals are set in a ground-mass of quartz, feldspar and glass.

BELOW RIGHT **View B under crossed polarized light** Under crossed polarized light, the quartz phenocrysts, which are different coloured greys because of their different orientations, take on a grey colour. The glass crystals are black.

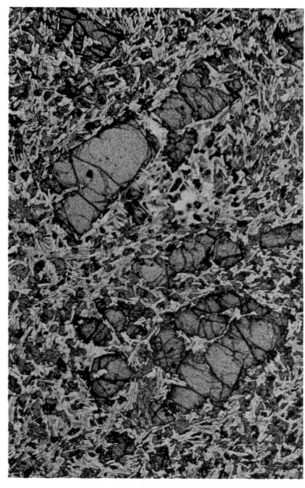

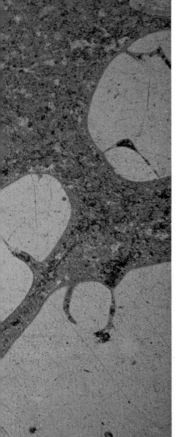

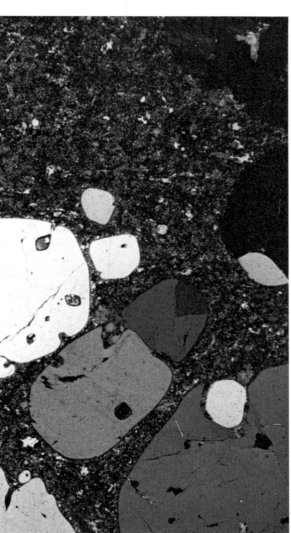

which occasionally proves useful in the identification of minerals. Some minerals dissolve in weak acid, others effervesce when a spot of dilute acid (generally hydrochloric acid) is dropped on them. Such minerals include calcite, malachite and azurite. The mineral siderite reacts with heated dilute hydrochloric acid.

Not all the tests described in this chapter need be applied in the diagnosis of a particular mineral. In many instances, three or four tests should be sufficient to arrive at an identification.

Less common minerals are more difficult to identify. There are a number of highly sophisticated scientific techniques used in the identification of such samples. For example, very thin slices of the mineral can be cut and examined under special petrological microscopes. Hand specimens of minerals have certain properties and characteristics, and in the same way, thin slices of minerals have sets of distinguishing characteristics. By means of special lenses and filters, a number of different images of the mineral composition can be seen, and the content thus identified.

Modern identification techniques also include the use of X-rays. These are particularly useful in analyzing the internal structure of crystal samples. The X-rays are reflected from the atoms inside the crystal and form a pattern on a photographic plate. The pattern can then be measured and analyzed.

Another well-known identification technique is to heat the crushed powder of a mineral, mixed with a catalyst, by means of a blow pipe. Certain minerals undergo colour changes at certain temperatures. If the heating process is continued, the mineral will eventually melt. The measurement of the melting point is another useful laboratory aid to identification.

COMMON MINERALS WITH METALLIC LUSTRE

Chart A

Hardness		Streak	Colour	Diagnostic Properties	Mineral
Scratched by finger-nail and knife (hardness of 2 or less)	1	shining black	grey-black	Marks paper like pencil, soils finger and feels cold. Greasy feel. Light for metallic mineral (s.g.2) Generally massive	Graphite
	1	Some weathered haematite and limonite – see below			
Scratched by knife (hardness 2–4)	2½	black (sooty-black)	silver-grey	Cubic crystals with cubic cleavage. Exceptionally heavy (s.g. 7½). Colour and streak	Galena (lead glance)
	3	black (grey-black)	red-brown	Massive, tarnishes to iridescent purples	Bornite (peacock ore)
	3½	black (greenish-black)	golden-yellow	Massive; iridescent tarnish. Crumbles when cut with knife, brittle	Chalcopyrite (copper pyrites)
	2½	golden-yellow	golden-yellow	Massive, in sheets or nuggets. Very heavy (s.g. over 8), malleable, hackly fracture, silver and copper tarnish	Gold
		silver white	silver white		Silver
		copper-red	copper-red		Copper
Cannot be scratched by knife, scratched by quartz (hardness 5½–7)	6½	black (greenish-black)	golden-yellow	Cubic crystals with striations. Massive or nodular. No cleavage. Darker colour when tarnished. Emits spark when struck by steel. Odour	Pyrite (iron pyrites; fools gold)
	6	black (greenish-black)	iron-black	Strongly magnetic. May have rusty surface	Magnetite (magnetic iron ore)
	6	cherry-red	reddish-brown; dark brown-black; greyish-red	Massive, often bulbous exterior. Streak	Haematite (Hematite; kidney iron ore)
	5½	yellow-brown	shades of yellow-brown	Massive, occasionally bulbous exterior or earthy masses	Limonite

MINERALS WITH NON-METALLIC LUSTRE (relatively soft)

Chart B

Hardness			Streak	Colour		Diagnostic Properties	Mineral	
Scratched by fingernail, coin or knife (hardness of 3 or less)		2½	white	black	cleavable	Massive or flakes which show good basal cleavage yielding thin flexible and elastic sheets	Biotite	Mica
			white	white	cleavable		Muscovite	Mica
		1	white	white to pale green (occasionally darker)	cleavable	In fibrous masses which yield thin flexible but non-elastic scales. Soapy feel, pearly lustre	Talc (soapstone)	
		2	white	shades of green	cleavable	Similar to talc but harder and vitreous lustre	Chlorite	
		2	white	white or colourless when pure (often tinted giving yellow, red or grey colourations)	cleavable	Occurs as crystals (selenite) fibrous (satin spar) massive (alabaster); cleavage 3 but poor; yields thin flexible, non-elastic scales	Gypsum	
	sometimes by coin	2½	white		cleavable	Massive or cubic crystals with cubic cleavage. Salty taste, soluble in water	Halite (rock salt)	
	sometimes by coin	2	white		cleavable	Massive or cubic crystals with cubic cleavage. Bitter taste, soluble in water	Sylvite	
	sometimes by coin	2	white	earthy-white to tinted yellow-red	cleavable	Earthy masses; adheres to dry tongue, plastic when wet. Earthy odour, greasy feel	Kaolinite (china clay)	
	sometimes by coin	2	white		no cleavage	Earthy egg-like masses	Bauxite	
		2	white	yellow	no cleavage	Massive or encrusting, colour diagnostic, waxy lustre, attracts paper when rubbed on clothes	Native Sulphur	
		2	yellow-red	orange to deep red	no cleavage	Massive with waxy lustre. Colour diagnostic	Realgar	

MINERALS WITH NON-METALLIC LUSTRE (mo

Hardness		Streak	Colour
Scratched by knife (hardness 3–5)	3		colourless or w (occasionally tinted)
	3		
	4		(purple, yellow
	4	buff (light to dark brown when weathered)	grey (brown shades when weathered)
	4	pale brown to pale yellow	brown common (grey, yellow, green, red or black)
	3	scarlet	scarlet
	3½	blue	blue
	3½	green	green
	3	pale green	green with yell & red-browns

MINERALS WITH NON-METALLIC LUSTRE (har

Hardness		Streak		Colour
Scratched by Quartz (hardness 5–7)	5½	white		play of colours on milky or dar background
	6	white		white
	6	white		Pinkish green- amazonstone* pearly-moonsto
	5½	white	bluish	dark blue (fleck with white, gol
	6	white	bluish	bluish-green
	4–6½	white		pale blue
	6½	white		grey-yellow to browns
	5½	white		yellow-green; brownish
	6½	white		green shades
	6½	white		green shades
	6½	grey		yellow-greens
	6	grey-green		dark green to black
	6	grey-green to brown		black
	6½	grey		black to brown

Diagnostic Properties	Mineral
Crystals or massive, rhombic cleavage. Effervesces in cold weak acid. Double refraction	Calcite
Plate-like masses commonest. Rhombic cleavage. Unusually heavy (s.g. $4\frac{1}{2}$)	Barytes (Barite; heavy spar)
Cubic crystals, often twinned. Octahedral cleavage. Crystal form and glassy lustre characteristic	Fluorite (Fluorspar; Blue John)
Massive or granular. Effervesces with hot acid. Magnetic on heating. Heavy (s.g.4), rhombic cleavage	Siderite
Coarse granular masses. Crystals with dodecahedral cleavages; resinous lustre	Sphalerite (zinc blende false Galena)
Usually massive; streak and weight (s.g.8) are distinguishing features	Cinnabar
Associated together, generally massive with bulbous external surfaces; crystals rare. Colour and streak diagnostic. Effervesces in cold acid	Azurite
	Malachite
Massive or fibrous (asbestos). Soapy feel, waxy lustre	Serpentine (and Asbestos)

Diagnostic properties	Mineral	
Blending of colours. Light weight (s.g.2); resinous lustre, conchoidal fracture	Opal*	
Crystals in rock, recognized by prismatic cleavage at 90° and hardness. Plagioclase has striations due to twinning	Plagioclase	Feldspar
	Orthoclase	
Massive. Light weight (s.g.2)	Lapis lazuli*	
Granular masses, sometimes bulbous. Colour diagnostic	Turquoise*	
Fibrous masses or long bladed crystals. Hardness varies 4 along, 6 across; crystal	Kyanite	
Fibrous masses or long needle-like crystals	Sillimanite	
Massive or lozenge-shaped crystals with brilliant lustre and prismatic cleavage	Sphene	
Tough compact masses	Jade*	
Granular olive-green masses	Olivine (Peridot*)	
Massive and granular. Small crystals with unique colour characteristic	Epidote*	
Short fat 4-or-8-sided crystals. Prismatic cleavage at 90°	Augite	
Long 6-sided crystals. Prismatic cleavage at 124° and 56°. Splintery on broken surfaces	Hornblende	
Massive and granular with dark shiny grains. Sometimes bulbous or as crystals. Brilliant lustre and heavy (s.g.7). Light coloured streak, weight & lustre diagnostic	Cassiterite (tin stone)	

MINERALS WITH NON-METALLIC LUSTRE (extremely hard; all are gemstones) Chart E

Hardness and Streak		Colour	Diagnostic Properties	Mineral & Gem Varieties
No streak and not scratched by quartz but will scratch glass (hardness 7 or greater)	7	colourless or variously tinted	6-sided crystals often with horizontal striations; conchoidal fracture	Quartz purple—amethyst pink—rose quartz red—citrine yellow—carnelian banded—agate green—onyx dark green—bloodstone
	10		Crystals. Feels cold when first touched. Attracts paper when rubbed on silk	Diamond
	8		Attracts paper when rubbed on clothes. Single cleavage	Topaz
	8		Large 6-sided crystals	Beryl green—emerald greenish-blue—aquamarine pink—morganite brown—heliodor
	9	grey (tinted)	Hardness and barrel-shaped crystals diagnostic. Brilliant lustre	Corundum blue—sapphire red—ruby
	8	red-brown (blue-black)	8-sided diamond shaped crystals	Spinel deep red—spinel ruby red—Balas ruby yellow—Rubicelle
	$7\frac{1}{2}$	red-brown (paler)	4-sided pointed crystals with brilliant lustre	Zircon red—hyacinth colourless—jargoon
	7	red-browns (green; black)	12- or 24-sided crystals; conchoidal fracture	Garnet numerous gem names
	$7\frac{1}{2}$	red-browns	crystals in form of cross, with dull rough surface	Staurolite
	7	black to dark green or variously tinted	slender 3- (6- or 9-) sided crystals with vertical striations. Attracts paper when rubbed.	Tourmaline red-pink—Rubellite blue—Brazilian sapphire green—Brazilian emerald yellow—Ceylon peridot colourless—Achroite
	$8\frac{1}{2}$	green shades (yellow, brown)	flat striated crystals	Chrysoberyl green by day— red by artificial light—Alexandrite
	$7\frac{1}{2}$	bluish (grey-yellow)	colour & colour changes (pleochroism)	Cordierite

Classification of mineral properties These tables show the most straightforward diagnostic characteristics—hardness, streak and colour—together with the other properties which it is necessary to know to identify the most frequently found minerals.

Chapter 2
IGNEOUS ROCKS

Geologists consider that igneous rocks were the first rocks to have formed on the earth, for only after these have formed can the other two rock types – sedimentary and metamorphic – be developed. The oldest rocks on earth are therefore igneous rocks, but paradoxically, the newest rocks are also igneous! For igneous rocks are being formed at this very moment.

The word 'igneous' is derived from a Latin word meaning *fire*, which provides a clue to the origin of the rocks. They are formed by the cooling and solidifying of extremely hot molten rock. It is difficult to imagine temperatures so great that rock itself is molten, yet we have plenty of evidence to show that such temperatures exist below the surface of the earth. Many areas of the earth contain points of weakness in the crust, and sometimes the molten rock within the earth forces its way through these points of weakness and appears on the surface. The spectacular photograph opposite is of the new island Surtsey, which appeared a few years ago off the coast of Iceland in the North Atlantic Ocean. It shows hot molten rock – called magma – running along the earth's surface.

Once magma has escaped from the conditions of high temperature and pressure which exist beneath the surface, it begins to cool down. Eventually the flowing magma slows down, comes to a stop and solidifies. The molten magma has become solid igneous rock. In a way, the process is very similar to that in a steel-works, where molten metal is poured from a blast furnace, cools, and solidifies into metal ingots. The group of igneous rocks, therefore, are those rocks which were once hot and molten, but which are now cold and solid. They are generally hard, tough rocks consisting of a small number of minerals fused together. The type and character of igneous rocks is determined by the composition of the original magma from which it was formed. It is usual to refer to three main types of magma, each type producing one of the three fundamental igneous rock groups.

One magma is rich in silicon and aluminium – quartz-making and feldspar-making materials – so that igneous rock formed from it will have an abundance of quartz and feldspar minerals present. Such a magma is known as an acid magma, and the resulting rock as acid igneous rock. Another magma type contains feldspar-making material, but has little or no quartz-making material. On cooling and solidification, it produces igneous rock composed largely of feldspar, with only small amounts of quartz. This type of magma is known as intermediate magma, and the resulting rock as intermediate igneous rock.

Surtsey, Iceland Molten rock (magma) erupting from this newly exposed volcanic island in the North Atlantic Ocean. Parts of the island – the dark outline seen in the picture – represent already cooled magma which has solidified into volcanic basalt rock lava. Surtsey is still in the process of formation.

CLASSIFICATION OF IGNEOUS ROCKS

	Original Magma Type			
Grain Size	**Acid**	**Intermediate**	**Basic**	**Cooling Location**
Fine-grained Individual minerals difficult to see	Obsidian Rhyolite	Trachyte Andesite	Basalt	Extrusive (on earth's surface)
1 mm				
Medium-grained Individual minerals generally seen	Quartz Porphyry	Microsyenite Microdiorite	Dolerite	Hypabyssal (shallow depths) — Intrusive (within the earth)
5 mm				
Coarse-grained Individual minerals clearly seen	Granite	Syenite Diorite	Gabbro	Plutonic (great depths) — Intrusive (within the earth)
Dominant minerals present	Quartz	Feldspars	Ferro-magnesian minerals	

ABOVE **Basalt hand specimen** A fine-grained, dark-coloured rock which has resulted from the cooling and solidification of basic magma. The dark spots are infilled gas bubbles, known as amygdaloidal basalt.

LEFT **Classification of igneous rocks** The original magma type and its cooling location are reflected in the grain size and the dominant minerals present in each of the three types of igneous rocks.

The third type, basic magma, is rich in iron and
magnesium. These are the ferro-magnesian mineral-
making materials. Such minerals are usually black
or dark green in colour, and include augite, olivine and
hornblende. Rocks formed from this type of magma
are known as basic igneous rocks.

Not only do the types of original magma result in a
considerable variation in mineral content, but the size
and shape of the mineral crystals is dependent on the
location within the earth's crust at which cooling and
solidification occurs.

Geologists generally consider two main locations;
magma may reach the surface of the earth in its molten
condition (as it does on the island of Surtsey, for
example); or it may solidify within the earth, before it
reaches the surface. Once it has reached the surface,
the magma cools quickly. The minerals contained in
it do not have time to develop properly, and so the
crystals formed are generally quite small. These
small crystals are difficult to see, even with the aid of
a hand lens, although they can be seen under more
powerful magnification. Such fine-grained rocks are
called *extrusive*, since the magma has been extruded on
the earth's surface, and solidified there. Açid, basic
and intermediate extrusive rocks may be developed,
depending on the original magma type.

Magma which becomes solid before it has a chance
to reach the surface becomes entombed below ground
level, and the igneous rock formed from it is trapped
between layers of other rocks. Below ground level the
magma cools at a much slower rate, so that the mineral
crystals have time to develop into a recognizable size
and shape. Such large-grained rocks are called *intrusive* rocks, since the magma has been intruded into
the earth's layers. Acid, basic and intermediate
intrusive rocks may be developed, again depending on
the original magma type. The size of the crystals formed
during the cooling process largely depends upon the
rate of cooling of the magma. Magma cooled very
slowly at great depths forms coarse-grained igneous
rocks, while magma cooled less slowly at higher levels
forms medium-grained rocks. Thus intrusive igneous
rocks can be further sub-divided into *hypabyssal* and
plutonic types – plutonic referring to those rocks
formed at great depths, hypabyssal to those formed at
shallow depths. Thus location, cooling history and
magma origin provide convenient data for the classification of igneous rocks (see table opposite).

The reasons for the existence of the three different
magma types are very complicated, and not yet fully
understood by geologists. In the simplest terms, it
appears that the magma types depend on the condi-

tions, deep inside the earth, under which they were first formed. Acid magmas are thought to originate by melting at the bottom of the earth's crust, at the base of the mountain masses. Basic magmas are thought to be the result of melting of the mantle layers. Intermediate magmas seem to originate along the downward edges of plate slopes.

Once these magmas have been formed, they make their way upwards, some to reach the surface of the earth, the majority to become trapped at some point below the surface. In either event, the magma gives birth to features which are often referred to as igneous *forms* or *bodies*.

Extrusive igneous bodies

Magma is being released onto the earth's surface at this very moment, and provides impressive evidence that molten rock can and does exist! It escapes either through a hole in the ground, or through a series of long thin cracks. In the first instance, the magma gradually builds up into a mound shape known as a *volcano*. In the second instance, it either builds a row of volcanoes, or solidifies as a large expanse of flat magma – a lava flow. The photograph on page 39 shows the first stages in the building of the Surtsey volcano off Iceland in 1963; more recently another volcano has started to form in the same area. Contrast this photograph with that of Mount Fuji below. This mountain is considered to be a perfect example of the final conical shape of a typical volcano. No activity has been recorded in Mount Fuji since 1707. As a result geologists say that it is *dormant* (sleeping) or *extinct* (dead), depending on whether they think any future activity is likely. The prediction of future events regarding volcanic activity is an uncertain occupation! At 9.56 a.m. (local time) on August 27, 1883, an explosion estimated to have been 26 times the greatest H-bomb test detonation blasted the island of Krakatoa. Rocks were thrown 34 miles into the air, dust fell 3000 miles away, the explosion was heard over a thirteenth part of the surface of the earth, and more than 36,000 people died. Many people had considered Krakatoa to be extinct!

Geologists estimate that there are some 450 known active volcanoes in the world with perhaps another eighty active submarine volcanoes. Recent geological

experiments have indicated that the earth's crust consists of a number of great plates – tectonic plates. Most of the active volcanoes are situated along the areas where these plates meet. This is reasonable because these meeting areas are point of weakness in the solid crust, and it is relatively easy for the magma to escape from the interior of the earth (see page 13).

Extrusive igneous rocks which have cooled from magma at the surface are known by the general term of *lava*. Large amounts of gas are always released during magma extrusion – the characteristic smoke of active volcanoes. This gas plays an important and interesting part in the formation processes of igneous rocks. When gas is released during volcanic eruptions, it causes the lava to break up into fragments, known as *pyroclastic* material. The fragments may be in the form of a fine dust or ash, or in the form of large boulders. Volcanic dust may form very thick deposits in the vicinity of a

ABOVE **Mount Etna, Sicily**
An active volcano in the Mediterranean; a slow but steady migration of liquid rock is seen escaping from the side of the mountain during Etna's most recent eruption in 1971.

LEFT **Mount Fuji, Japan**
This mountain shows a typical cone-shaped composite volcanic form. Its first eruption was recorded in 781 and there have been about ten eruptions since then. The last occurred in 1707, and since then Mount Fuji has been dormant.

ABOVE **Mount Erebus, Ross Island, Antarctica** Mount Erebus shows the typical smooth dome of a volcano of the composite type, where layers of ash and lava alternate to create a volcanic mould.

LEFT **Basalt columns, Bavaria** Fairly rapid cooling of magma on the surface often results in the formation of hexagonal pillars at the base of the lava flow.

BELOW **Tuff deposits, Canary Islands** Volcanoes which are highly explosive often release huge quantities of dust and ash into the air. These settle and form thick pyroclastic layers of ash, which are known as tuff deposits.

volcanic eruption. When these deposits become consolidated and harden, they are known as *tuff*. Coarse fragments, too, may form thick deposits. Loose deposits are known as *cinder*, consolidated deposits are known as *agglomerate*. Coarse fragments often include volcanic bombs, which are once-hot pieces of molten rock, moulded into spindle-shapes during their flight through the air.

There are three basic types of volcano, classified according to the material from which they are built. *Composite* cones have alternating layers of ash and lava, and are the commonest type. Mount Fujiyama, Mount Erebus and Mount Etna are all examples of composite cones. *Lava* cones, as their name implies, are composed mainly of lava. Some are very low but extensive mounds, and are often called shield volcanoes because from the air they resemble ancient warrior shields. Others are high, steep-sided mounds. *Cinder* cones are

RIGHT **Lava flow and volcano, Canary Islands** The now dormant volcano can be seen in the background. The foreground is composed of a rough lava field and indicates the once-active nature of the volcano. Note the blocky nature of the lava ('aa' lava) caused by the breaking up of the cooling, but still flowing, lava.

BELOW **Lava drips from Mount Etna** Intricately interwoven dribbles of lava often result when lava drips from a lava flow. Sometimes they occur when hot pieces of magma are thrown into the air, twisting and hardening as they fly. This specimen is three inches long.

composed mainly of ash and cinders. Cerro Negro in Nicaragua is a good example of this type of volcano.

Volcanic activity often results in weird and spectacular rock formations. The photograph on page 45 shows such a feature. It is known as a volcanic neck and is situated at Le Puy in the Auvergne region of central France. A volcanic neck is produced as the hardened remnants of lava from the throat of a volcano core are left as the surrounding volcano weathers away over the ages. Such hard stumps of rock can be seen at Shiprock (New Mexico), Matchett's Pyramid (N.W. Australia) and Castle Rock (Edinburgh).

A lava flow, or if it is extensive, a lava plateau, is a familiar feature of extrusive igneous activity. The photograph on the right illustrates some of the characteristics which are typical of lava flow: a nineteenth-century lava flow in the Canary Islands shows the 'blocky' surface which is known as *aa* (a Polynesian word, pronounced 'ahah'.) Other flows of the same period may show the rope-like surface known as *pahoehoe* (a Polynesian word, pronounced 'pahoy-hoy'). Lava flows often contain rock which is riddled with holes. Such lava is known as *vesicular* lava, and the holes have been produced by the escaping gas during cooling. In rare instances the holes have been infilled with gemstones such as amethyst, opal and agate; however the less exotic mineral calcite is more often found as infilling material. Although it cannot be clearly seen in the photograph, vesicular lava is feature of the Canary Islands flow.

One of the most spectacular and exciting features of lava flows are the giant columns often formed. The Giant's Causeway, Northern Ireland, and Fingal's Cave, Scotland are fine examples of this phenomenon. The columns of basalt which make up the Giant's Causeway were formed by the tremendous forces which are operative during the cooling of volcanic lava. The cross-section of these columns is hexagonal, and the columns formed at right angles to the surface of the lava flow. Millions of years of weathering have broken down quantities of the lava into screes, largely consisting of hexagonal 'cakes' of rock. Yet another unusual feature of lava flow results when lava is erupted into water, as may happen if the volcanic activity is along a coast or beneath the sea. The lava has the appearance of a number of pillows, piled on top of each other, and therefore is known as *pillow lava* (see page 47).

Gas is continuously released during volcanic activity. During the final stages of activity no magma is released, although gases are still emitted. Carbon dioxide is released as a late product in areas such as Death Valley in Java, and Death Gulch in Arizona – sinister names suggesting the effect these volcanic gases have had on the animal population. Hot springs, geysers and boiling mud pools are also characteristic features of volcanic areas. Water at depth becomes heated by the still hot volcanic rock. This heated water sometimes reaches the surface and lies in pools or lakes. Sometimes it is forced to the surface through narrow

LEFT **Volcanic ash, Thera** A close-up of fine white ash. The position and thickness of earlier layers of lava are clearly seen in the background. The ash layer varies in thickness from 30 ft to 100 ft; here it is about 60 ft thick.

BELOW **Vesicular lava, Thera** A close-up of vesicular lava. The honeycomb texture is caused by gas bubbles within the cooling magma.

fissures and under pressure, and forms geysers. Old Faithful in Yellowstone National Park, United States, is probably the most famous example of a geyser, but the name 'geyser' comes from the Great Geyser, in Iceland, which is no longer active. Geysers and hot springs often contain quantities of dissolved mineral material, which is deposited in mounds around their edges.

Intrusive igneous bodies

Spectacular (and sometimes disastrous) as volcanoes and lava flows are, they represent only the magma which has managed to reach the earth's surface. The majority of the molten rock material never reaches the surface, but cools and solidifies within the earth's outer layering.

ABOVE **Giant's Causeway, Co. Antrim, Northern Ireland** These hexagonal columns were formed at the base of a lava flow. The Giant's Causeway was once part of an extensive lava field, relics of which occur in the Western Isles. The joining of the columns is due to uniform cooling.

LEFT **Le Puy, Auvergne, France** The tall remnant of the neck of a volcano after erosion has removed the soft surrounding rock neck.

45

to the layering — the angle varies, but they can sometimes be at right angles — they are known as dykes. The layering in which sills and dykes occur is generally known as a bedding plane and is a characteristic of sedimentary rocks (see Chapter 3). Sills usually form quite near the surface of the earth, and so are generally composed of medium-grained hypabyssal rocks. A good example is the Great Whin sill in northern England, which is composed of dolerite. The Palisade sill in New York and the Karoo sills in South Africa are also composed of dolerite.

Other sills, however, may be formed at deeper levels and so consist of coarse-grained plutonic rocks. The Salisbury Crag sill in Edinburgh and the Nahant sill in

ABOVE **Hot spring, Yellowstone National Park** Typical hot springs are considered to be the final stage of volcanic activity. The highly sulphurous landscape is a characteristic feature.

Magma that solidifies from the molten stage within these outer layers also forms recognizable igneous bodies. But unlike extrusive activity, the intrusive processes cannot be watched directly and geologists have to wait until erosion wears away the overlying rocks before they can study the igneous features below. As a result, the intrusive igneous bodies studied by modern geologists are all very old — sometimes formed many millions of years ago.

Of the common intrusive bodies, only *sills* and *dykes* are clearly recognizable as such at the surface (once the rock covering has been worn down sufficiently to reveal them). *Batholiths*, another common intrusive body, are less recognizable. Generally, erosion only reveals the top of the batholith and the geologist must use other evidence to determine its nature. Less common intrusive bodies, recognized at the surface after erosion, are *ring dykes* and *cone sheets*. The intriguing names *laccolith* and *lopolith* describe other igneous bodies which are sometimes so large that usually only the tops are exposed after thousands of years of erosion.

Sills and dykes

Sills and dykes are sheets of igneous rock which have been intruded into the earth's covering layers of rocks. If the sheets run parallel to the layering they are known as sills. If the layers are horizontal, then the sill is also horizontal. If the layers are tilted, then the sill is inclined at the same angle. If the sheets are not parallel

Massachusetts are both composed of gabbro, while the Shonkin sill in Montana is composed of syenite. The photograph above shows another syenite sill, on the east coast of Nova Scotia, Canada. The plutonic syenite has intruded into a sedimentary shale which was laid down during the Palaeozoic Era.

Sills, like most intrusive bodies, are normally exposed by a downward wearing-away of the rocks above them. The Great Whin sill has been exposed in this way, for example. In this instance, the erosional process has continued so that the sill now protrudes above the ground, since the sedimentary rocks into which it was intruded were softer and more easily worn away. In contrast, the North West Territories sill in Canada has been exposed by sideways erosion. This means that geologists are able to study its exact position within the crust, for the rocks above it, which normally would have been removed by erosion, are still in position. The same sill illustrates another important sill feature — transgression. Transgression takes place when there is a change in the run of the sill, and it 'jumps' from one bedding plane to another. Such a sill is said to be transgressive, because it transgresses (changes) horizons. Most known sills, including the Great Whin sill, have this feature, but because of the erosion patterns it is often difficult to appreciate with the naked eye. Close study of a geological map, or of a reconstructed cross-section, reveals the feature clearly.

Sills vary in thickness. The Karoo sill is thousands of feet thick in parts, the Palisade sill is about one thou-

RIGHT Dyke in Devonian slate, Cornwall The discordant arrangement of the dyke to the bedding planes of Devonian slate is clearly seen.

ABOVE Syenite sill, East Nova Scotia Magma which fails to reach the earth's surface often becomes emplaced near the surface, having squeezed itself along bedding planes of sedimentary rock. This syenite sill shows the parallel arrangement with the sedimentary rocks.

BELOW Pillow lava, Ballantrae When liquid magma escapes and either flows into the sea or is released under the sea, it is so quickly quenched that cushion-like balls of lava result.

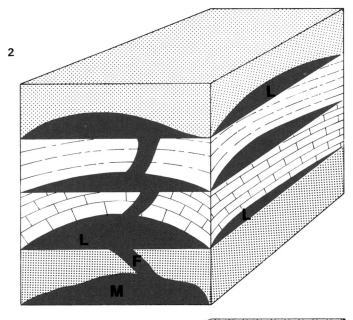

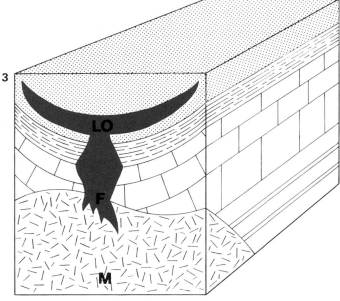

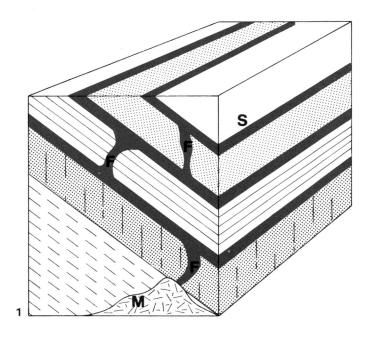

A series of drawings to show the formation of various features associated with igneous rocks.

1 Sill Sills (s) are formed when magma (M) is injected along bedding planes present in the rock strata, cools and solidifies. If the strata are tilted, the sill will be inclined at the same angle. F indicates faults.

2 Laccolith Laccoliths (L), usually formed at depth, occur when magma movement is restricted along the length of the bedding plane. The magma (M) tends to lift the strata and to solidify in mushroom-shaped mounds.

3 Lopolith A lopolith (LO) is a basin-shaped igneous rock body probably caused by the subsidence of the crust followed by a magma intrusion. Lopoliths, like laccoliths, are generally found to be deep-seated igneous bodies.

4 Dyke Magma which is forcing its way towards the earth's surface, regardless of the strata above it, often cuts across pre-existing bedding planes. When this happens, the magma cools and forms the igneous body known as a dyke (D).

sand feet thick, while the Great Whin sill is only about one hundred feet thick. The ages of sills vary a great deal, too. The North West Territories sill is Pre-Cambrian, and is at least 1000 million years old. The Nahant sill is about 450 million years old, and the Great Whin sill about 300 million years old. The Karoo and the Shonkin sills are younger, being formed during the Mesozoic Era. The Karoo sill is about 200 million, the Shonkin about 100 million years old.

Dykes are generally much thinner sheets of molten rock which have cut across the pre-existing bedding plane of sedimentary rock and which have solidified in that position. The diagram above and the photograph on page 47 show the relationship of the dyke to the bedding planes. Dykes, like sills, usually form near the surface, so they are generally composed of medium-grained hypabyssal igneous rocks. Coarser-grained rocks are rare, because the narrowness of the dyke allows a rapid heat loss. Even if the dyke were formed at depth, this heat loss would allow little chance for large mineral crystals to form.

Batholiths

When large amounts of magma accumulate and solidify at great depths within the bases of mountain chains, coarse-grained plutonic bodies develop. These enormous bodies of igneous rocks are known as batholiths, which, appropriately enough, means 'deep rock'. The magma has solidified at depths greater than any other intrusive body, so that a long period of erosion is needed before they are exposed. Even after millions of years, only the tops of these huge structures can be seen (see page 50).

The gases which escape into the air during extrusive igneous activity are still present at depth, but may escape, together with some liquid magma, into the surrounding rocks. Depending on the chemical composition of the escaping mixture of gases and liquid

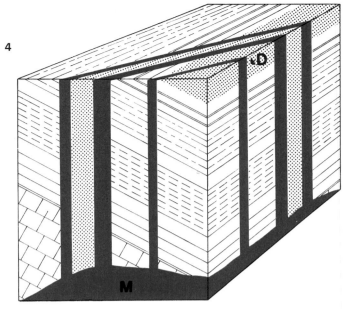

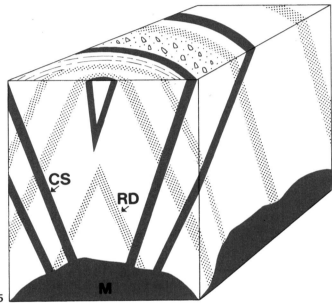

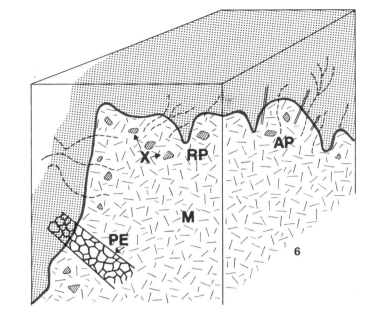

5 Cone sheets and ring dykes Cone sheets (CS) are steeply inclined sheets of rock filling cone-shaped fissures in the earth; the fissures are probably formed by upward pressures. The point of the cone is downwards. Ring dykes (RD) appear as a series of inverted cones, probably the result of subsidence followed by infilling of the fissures with magma.

6 Batholith Batholiths are irregular bodies of igneous rock, mostly of granite: they are of unknown depth and many thousands of square miles in areal extent. They occupy the cores of the great mountain chains. Their mode of origin is uncertain. Batholiths are exposed at the surface of the earth by erosion. Dykes of fine grained rock (aplite veins: AP) and coarsely crystalline rock (pegmatites: PE) are produced in the cooling. Pieces of undigested surrounding rock (xenoliths: X) and hanging pieces of the original rock (pendants: RP) are also features of batholiths.

magma, quartz veins, ore-mineral veins, aplite veins or pegmatite veins may be formed on cooling. Aplite and pegmatite veins are often formed within the body of the parent igneous mass. These are a feature often associated with batholiths. The gases and liquids are instrumental in changing both the igneous rock and the surrounding rock. In the latter event metamorphic rocks are sometimes formed. The tremendous heat emitted from the vast molten mass of the embryo batholith bakes the surrounding areas of rock, and this also results in the formation of metamorphic rocks (see Chapter 5). The association of ore-mineral veins with intrusive igneous bodies is important to the miner. When prospecting for ores, mining companies often concentrate their searches in areas where batholiths occur.

Batholiths often reach an incredible size. In fact, any large exposure of granite or other coarse-grained igneous rock is probably part of an exposed batholith.

In Britain the granite masses of Dartmoor, Bodmin, Land's End and the Scilly Isles are all small exposed parts of the same huge batholith. The Coast Range, Idaho and the Sierra Nevada and Patagonian granite masses are all probably part of one gigantic batholith which stretches, deep underground, from the top of North America to the tip of South America. Batholiths are also important because they often provide the reservoirs for the other intrusive bodies and for surface magma.

Less common igneous bodies include *laccoliths* (from the Greek *lakkos*, a reservoir, and *lithos*, rock), *lopoliths* (from the Greek *loppos*, a basin or flat earthen bowl), *ring dykes*, and *cone sheets*.

Laccoliths are mushroom-shaped domes, generally composed of medium-grained or coarse-grained rocks. They have the same relationship to the bedding planes as have sills, and are probably locally thickened sills which have failed to squeeze along the bedding plane, remaining in one position and arching over the rocks above. Good examples of these igneous bodies are found in the Henry Mountains in Utah and in Judith Mountain, Montana, in the United States. Sometimes a number of laccoliths are formed from one central source, and lie one above the other. Such formations are described as 'cedar-tree laccoliths', and good examples can be found in the Builth Wells area of Central Wales.

Lopoliths resemble huge saucers, thickened towards the centre. They may have been formed by the collapse of the subsurface, probably caused by disturbances in the earth's crust; or by the withdrawal of a body of magma followed by renewed intrusion. Lopoliths often show a layered structure, which may be the result of gravitational sinking of the early-formed crystals. For example, the Insizwa lopolith in South Africa grades downwards from a quartz gabbro, to gabbro and other basic rocks known as norite and picrite. The rock type and density change with depth, becoming more basic and more dense. It has been calculated that the volume of the Insizwa lopolith is 300 cubic miles. The Sudbury lopolith, in Ontario, Canada, is thirty-six miles long and seventeen miles wide.

Phacoliths are another type of intrusion, and are located in the crests and troughs of intensely folded rocks. Phacoliths are often potential ore-carriers, as in the gold-bearing quartz phacoliths of the Bendigo Mine in Australia. Ring dykes and cone sheets are steeply inclined dyke-like intrusions, giving a distinctive oval outcrop pattern. They are formed when magma squeezes up cracks developed as a consequence of great pressures set up by the intrusion of a magma reservoir at an even greater depth. Ring dykes and cone sheets may occur as single oval fractures, as in the ring dykes of Mull in Scotland, Askja in Iceland and

BELOW **Obsidian outcrop, Lipari Islands** The black, glassy nature of obsidian with, in this example, white gas bubble streaks, is the result of the very rapid cooling of the magma.

BELOW RIGHT **Orbicular diorite, Finland** Occasionally, as a result of cooling, minerals separate out into layers; the dark (ferromagnesian) layers and the white (mainly feldspar) layers in this example are generally represented by irregular scatterings through the rock in other non-orbicular igneous rocks.

LEFT **Granite batholith, Yosemite National Park** Only the top part of the deep-seated intrusive body known as a batholith is actually seen in this photograph. Batholiths are often quite massive. They are usually composed of granite, and are formed when huge underground reservoirs of magma solidify before the magma can escape (see page 49).

BELOW **Xenolith in granite, Yosemite National Park** Batholiths are also characterized by pieces of surrounding rock which give the impression of having fallen into the granite magma melt. More probably, the granite 'chewed' its way into the rock, melting most of it, and leaving chunks of undigested rock. These are known as xenoliths.

Medicine Lake in California; or as a series of concentric rings, as in the Oslo ring complex in Norway. The unusual occurrence of both ring dykes and cone sheets at one locality can be seen at Ardnamurchan Point in Scotland, the most westerly point of mainland Britain.

Minerals in igneous rocks

Silicon and oxygen are the most abundant elements found in igneous rocks, and so most of the minerals are either oxides or silicates of various other elements. Aluminum, iron and magnesium are important ingredients of the darker and denser minerals – the ferro-magnesian minerals. Lime is important in the formation of some feldspars, while potassium and sodium are ingredients in other feldspars. The total number of materials which go into the formation of igneous rocks is small, but minute differences in their distribution between members of the same mineral family give rise to variations in the composition of the rocks. Thus there are many more types than would at first appear possible. The minerals which compose igneous rocks can be divided into three groups:

The essential minerals, which determine the character of the rock and which constitute its greatest volume. These minerals are the feldspars, quartz, hornblende, augite, the micas and olivine.

The accessory minerals, which usually only form a small part of the total volume of the rock. Accessory minerals are often useful to the geologist because they provide evidence of the physical and chemical conditions under which the rocks were formed. Some common minerals in this group are rutile, apatite, tourmaline, zircon and topaz.

The secondary minerals, which are formed by alteration of essential and accessory minerals. The clay minerals and chlorite are examples of secondary minerals.

The essential minerals

1. The feldspar family There are four basic molecules, (or 'building units') in this family. Each may mix with combinations of the others, so that a wide variety of feldspar minerals are formed. The four molecules are *orthoclase* ($KAlSi_3O_8$ – a chemical shorthand that tells us that the elements potassium (K), aluminium (Al), silicon (Si) and oxygen (O) are present); *albite* $NaAlSi_3O_8$ – sodium (Na), aluminium, silicon and oxygen are present); *anorthite* ($CaAlSi_2O_8$ – calcium (Ca) is present); and *celsian* ($BaAl_2Si_2O_8$ – barium (Ba) is present). The most important groups of this large feldspar family are the potash feldspars, with orthoclase as the most important ingredient, and the soda lime feldspars. The potash feldspars are essential minerals in the granite and syenite families of igneous rocks. The soda lime feldspars also occur in the granite and syenite families, with albite as an essential ingredient. They also occur in diorite and gabbro – rocks of the intermediate and basic igneous groups. Anor-

thite is an essential mineral in basic rocks such as anorthosite.

2. **The quartz family** Quartz (SiO_2) is an essential mineral in acid igneous rocks; almost an accessory in the intermediate group; and absent in the basic group. The presence of large quantities of quartz is conclusive evidence that the rock in question belongs to the acid group.

3. **The amphibole family** This family consists of the silicates of magnesium, iron, calcium, sodium and potassium. Hornblende, a complex mineral containing calcium, sodium, magnesium, iron, aluminium, silicon, oxygen and hydrogen atoms, is one of the common minerals in this family, and is an essential ingredient in the intermediate group of igneous rocks. Asbestos is a metamorphic material (see Chapter 7) and is another important member of this family, its insulating properties making it very useful to industry.

4. **The pyroxene family** This family comprises sili-
cates of magnesium and calcium, with aluminium appearing occasionally and sodium and lithium rarely. Augite is an essential ingredient of gabbro.

5. **The mica family** Biotite and muscovite are the two most common members of the mica family. They are the silicates of aluminium, potassium, iron and magnesium – the latter two elements contributing to the dark colouration of biotite. Sodium, titanium and lithium also appear in the 'family tree'. The micas are essential ingredients of the acid rocks such as granite. They also form giant crystals in the pegmatites found in Madagascar and India.

6. **The olivine family** Forsterite and fayalite are two members of the olivine family, but the most important member is olivine itself. This is essential to the basic gabbros, and in the peridotites. The presence of olivine is evidence that the rocks in question are basic, since the presence of greater quantities of silicon would result in the formation of other ferro-magnesian

ABOVE **Granite hand specimen** The coarse grain and light colour of this rock are typical features of an acid igneous rock. The white minerals are feldspar, the grey patches quartz minerals and the black minerals biotite mica.

RIGHT **Syenite hand specimen** A pale, coarse-grained rock of the intermediate igneous group of rocks. At first sight, syenite can be confused with granite, but the minerals here are feldspar (white) and hornblende (black). No quartz minerals are seen although a small amount of quartz would constitute the smaller mineral grains.

BELOW RIGHT **Syenite thin section** This shows the fused nature of igneous rocks. The large white crystals are orthoclase feldspar which dominate the mineral. Most of the smaller crystals are of orthoclase with small amounts of quartz, plagioclase feldspar and ferro-magnesian minerals.

BELOW **Granite thin section** Thin sections are often the best way to see clearly the minerals which make up a rock. This thin section shows feldspar (white), quartz (grey) and mica (black) minerals. Sections are photographed using a polarizing microscope, and identification of specimen is based on colour changes which take place as the section is rotated.

materials found in intermediate rocks. The green semi-precious stone *peridot* is a member of the olivine family.

The accessory minerals

Important examples of this group are rutile (TiO_2), which is an oxide of titanium, zircon ($ZrSiO_4$), tourmaline, which is a complex boro-silicate of aluminium, topaz ($Al_2F_2SiO_4$), and apatite ($Ca_5F(PO_4)_3$). Rutile and zircon are found in acid igneous rocks. Tourmaline and topaz are found along the margins of granite intrusions. Apatite is found in a variety of igneous rocks, but is particularly important in certain rocks of the gabbro family.

The secondary minerals

These minerals are the result of chemical changes in early formed essential and accessory minerals. The feldspars yield clay minerals such as kaolin and

53

montmorillonite; the biotite micas yield chlorite; the olivines, serpentine. The latter mineral forms the mass of rock serpentine in the Lizard Peninsula in England, and the great serpentine belt of Australia.

The acid igneous rocks

Acid igneous rocks are formed from magma rich in silicon and aluminium – the quartz-making and feldspar-making materials. Such rocks are therefore rich in minerals belonging to the quartz and feldspar families. Acid rocks can be either extrusive or in-intrusive, and therefore their grain size varies from

very fine to coarse, depending on the cooling location of the rocks.

Granite, the most common of the intrusive plutonic rocks, is a light coloured, coarse-grained rock. The colouration is due to the presence of large and well-formed feldspar and quartz minerals. Granite ranges in colour from the almost pure white of Oporto (Portugal) and Singapore granites, where white plagioclase feldspars dominate, to the grey granite of Dalbeattie in Scotland. Light pinkish granites, with both plagioclase and orthoclase feldspars, are found in situations as far removed from each other as these: Corsica, Hong Kong, Bornholm in Denmark and Shap

ABOVE **Granite, Yosemite National Park** Weathering and erosion have stripped off layer after layer of granite, rather like peeling layers off an onion. Such disfiguration is known as *exfoliation*.

which is located in the northern part of England.

Deeper red granites are found in Nystad in Finland, Stronstad in Sweden and Victoria in New South Wales. In these granites, pink orthoclase feldspars are the dominant mineral. In most of the granite rocks, flakes of biotite mica are present in small quantities, and this gives a textured appearance to an otherwise pale rock. Granite is a popular and valuable building material. Its hardness, strength and attractive appearance make it ideal for public buildings, banks and churches, although in granite areas whole towns may be built from locally quarried rock. Aberdeen, in Scotland, is a good example. Granite buildings are so predominant that the city is sometimes named 'the Granite City'.

Sometimes other minerals may develop in veins within the granites. This rock is called **pegmatite**, and it often contains large well-formed crystals of feldspar, quartz and mica. The gem minerals topaz, sapphire, ruby, emerald, beryl and aquamarine are found associated with pegmatites (see Chapter 6).

Quartz porphyry rocks have the same mineral content as granite, but the cooling history of the rock has allowed some of the minerals – sometimes quartz, sometimes feldspars – to develop large and recognizable forms. The other minerals remain small in size. The association of the different mineral sizes gives an attractive texture to the rock, a texture known as *porphyritic*. Collectors can find this rock type in sills and dykes associated with granite batholiths.

Acid magma which has reached the earth's surface has the same mineral composition as that forming the granite batholiths deep within the crust. However, the rapid cooling has resulted in the formation of small crystals, and the feldspar, quartz and mica minerals can only be recognized in thin section studies under a microscope (see page 35). The name given to this type of rock is **rhyolite**. **Obsidian** is an acid extrusive rock which has been cooled so rapidly that its minerals are indistinguishable, even under a microscope. Surprisingly, in this group of pale rocks, obsidian is jet black. The colour is due to the presence of tiny flakes of iron scattered throughout the rock. These can be seen with a hand lens. Obsidian is sometimes called volcanic glass, and when split has very sharp edges. In the past it was used to make arrow heads, axes and other cutting implements. Grey comet-like tails in the rock are the remnants of gas bubbles. Sometimes the bubbles are so numerous that a spongy, grey rock results. This is **pumice**, and is so light that it will float

BELOW **Porphyritic rhyolite View A under plane polarized light** This porphyritic rhyolite specimen shows euhedral quartz phenocryst in a ground mass of feldspar and biotite.

BOTTOM **View B under crossed polarized light** This shows the feldspars as grey dark. The needle-like brown crystals are biotite. The remaining small tabular crystals are quartz containing an iron oxide.

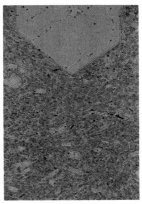

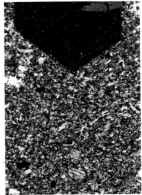

LEFT **Basalt, Madras** Boulders of basalt which have been weathered from the main outcrop are strewn around.

on water. Pumice occurs widely in the western states of the United States of America.

The intermediate igneous rocks

Syenite and diorite are the coarse-grained rocks of this type. Feldspars are the dominant minerals, the pink orthoclase feldspar giving a pink hue to syenite, the white plagioclase feldspar dominating diorite. Quartz minerals are generally absent, or present only in small quantities. Biotite and hornblende give the intermediate rocks a deeper grey appearance and help distinguish them from the granites. Syenites occur near Oslo in Norway, Dresden in Germany and Assynt in Scotland. Batholiths often have a diorite make-up.

Microsyenite and microdiorite are medium-grained intermediate rocks, while **trachyte** and **andesite** are fine-grained extrusive rocks. Trachyte is the common rock type of the Puy volcanic necks of the Auvergne region of central France. Andesite occurs in the volcanoes that encircle the Pacific Ocean.

The basic igneous rocks

The basic igneous rocks contain no quartz. They are usually dark in colour and heavy, both properties resulting from the presence of the ferro-magnesian minerals augite and olivine. The plutonic basic igneous rock is **gabbro**, and is often found in lopoliths and ring complexes (see page 48). **Peridotite** is a variety of gabbro and is very rich in olivine. Metal ores such as nickel, chromium and platinum are found associated with peridotite. In South Africa, diamonds are also associated with this rock. A medium-grained basic rock, **dolerite**, is similar to gabbro, the only difference being in grain size. A much finer grain size is a characteristic of **basalt**, although well developed augite or olivine crystals occasionally occur. Basalt is the most common rock on the earth's surface, since it escapes from submerged mid-oceanic ridges to form the sea floor. On land it forms huge lava flows such as the Deccan plateau in India, the Columbia-Snake plateau in the United States and the Western Victoria plateau in Australia. The largest basalt plateau is the Parana plateau in Brazil, which is 300,000 square miles in area. The escape of gas bubbles during formation results in *vesicular* basalt. Later infilling of these holes results in *amygdaloidal* basalt.

Recognition of igneous rocks

Active volcanoes and lava flows are obvious areas to obtain specimens of extrusive igneous rocks, and pyroclastic material can generally be collected in similar locations. Geologically older extrusive rocks are difficult to recognize without specialist knowledge

OPPOSITE **Microsyenite hand specimen** A medium-grained rock in which the dominant minerals are feldspar (grey), quartz (white) and hornblende (dark coloured).

OPPOSITE **Quartz porphyry hand specimen** Porphyry is a term used to describe igneous rock in which large recognizable minerals are set in a mass of finer materials. The larger minerals in this specimen (known as phenocrysts) are feldspar and quartz. They are set in a groundmass (known as the matrix) of finer minerals.

BELOW **Obsidian hand specimen** The conchoidal (shell-like) fracture is typical of obsidian.

or geological maps. Intrusive igneous rocks are easier to recognize. The mineral composition – well-shaped feldspar minerals, quartz, ferro-magnesian minerals – provides some evidence, as does the fused appearance and the hard tough character of the rocks. The location in which intrusive rocks are found – dykes, sills, batholiths, lapoliths – also provides good evidence for identification. These three clues – mineralogy, texture and location – are often sufficient to identify igneous rocks. An additional clue – igneous rocks never contain fossils. These are found in the sedimentary rocks discussed in the next chapter.

Both sedimentary and metamorphic rock formation takes place very slowly and in locations not accessible to geologists. But some igneous rocks are being formed on the earth's surface in places in which geologists can observe easily. Thus igneous rocks provide a unique opportunity for the study of rock formation. Knowledge gained through such a study allows geologists to obtain a deeper understanding of rock formation.

ABOVE **Gabbro hand specimen** The white minerals are plagioclase feldspar, the dark minerals are the ferro-magnesian minerals augite and olivine.

BELOW **Map showing the distribution of igneous rocks throughout the world**

Chapter 3
SEDIMENTARY ROCKS

In the last chapter it was stated that basalt was the most abundant rock on the earth's surface. So it is – but only because most of the ocean floor is made up of this igneous rock. But if we consider that part of the earth's surface which is above sea level – the dry land areas – then the most abundant rocks are those belonging to the sedimentary group. Sedimentary rock, as the name implies, has formed from layers of accumulated sediment. Some consist of the consolidated accumulation of material derived from the debris of already existing rocks as a result of various breakdown processes. Others may result from an accumulation of debris derived from organic material – the remains of plant and animal life.

There are three basic processes by which debris of different types can be accumulated. In the first, pre-existing rocks of any group – igneous, metamorphic or sedimentary – are broken down by the continuous process of weathering to form debris known as sediment. This process has been continuous over many millions of years and the early sediments were gradually buried as more and more sediment accumulated. The accumulated sediment is finally compacted through the weight of the sediment layers one above the other, and the cementing action of minerals taken into the sediment in solution in water.

In the second process, the sediment is an accumulation of plant and animal remains. The burial and compacting processes are similar to those yielding rocks belonging to the first group.

A third group of sedimentary rocks is formed by a process of mineral accumulation by precipitation from water which is saturated with a particular mineral.

Some sedimentary rocks have complex origins, where two or all three processes have occurred at the same time. As a result of this interplay between the formation processes, many interesting varieties of a particular type of sedimentary rock may occur. This adds to the interest of rock collection but complicates the problems of identification for the amateur geologist.

The complex origins of many sedimentary rocks also pose problems of classification, but fortunately, one of the three formation processes generally predominates. As in the charts of the igneous rocks (page 40) and metamorphic rocks (page 85), further information is necessary before a satisfactory classification can be made. In the debris-accumulated sedimentary rocks, the size of the debris pieces provides a useful distinguishing feature, for debris pieces can vary from the microscopic fragments which made up the silts and

Moab Deadhouse, Utah
Piles of sedimentary rock have been built up in sequence, and occur as flat horizontal sheets. The layers can be seen distinctly, and the flat tops of the 'mountains' are a feature of this area of Utah. Weathering and erosion have also played their part in sculpting out the landscape.

ABOVE **Shale hand specimen** Fine sediments such as clay, silt and mud are laid down to produce shale. Shale always shows minute layering.

BELOW **Classification of sedimentary rocks** Sedimentary rocks are classified first by their mode of origin and then by their content.

CLASSIFICATION OF SEDIMENTARY ROCKS

Sedimentary rocks originating as debris accumulates	Size of sediment pieces	Over 2mm Coarse	Conglomerate; Breccia
		2mm – 0.05mm Medium	Sandstone
		Less 0.05mm Fine	Shale; Siltstone; Mudstone
Sedimentary rocks originating as either plant or animal accumulates	Plant		Coal; Lignite; Jet
	Animal		Organic (shelly) limestone; Organic chert
Sedimentary rocks precipitated as mineral accumulates	Calcite		Limestone
	Iron		Ironstone
	Salt		Rock salt
	Gypsum		Gypsum rock
	Silica		Chert and Flint

muds, through the larger grains of the sands, to coarse, pebbly fragments of scree debris. Those sedimentary rocks which are formed as a result of mineral precipitation can be classified according to their mineral composition. For example, crystalline limestones are formed where calcium carbonate is precipitated. The third group of sedimentary rocks, those formed by the accumulation of plant and animal debris, can be classified according to their origin. For example, abundant plant remains form rocks belonging to the coal family.

The formation of sedimentary rocks is a continuous process. The sand on our beaches, the mud in our streams and rivers, and the scree debris on our mountain sides are the raw materials which will, millions of years in the future, form sedimentary rock. The processes of debris formation and accumulation can thus be studied as they happen. Because these processes are very much the same as those which formed existing sedimentary rocks millions of years ago, geologists can apply their observations to the past. Sedimentary rocks, therefore, provide a great deal of valuable and interesting information about the climatic and geological conditions prevailing in the area and at the time they were formed.

The amateur geologist or rock-collector can also

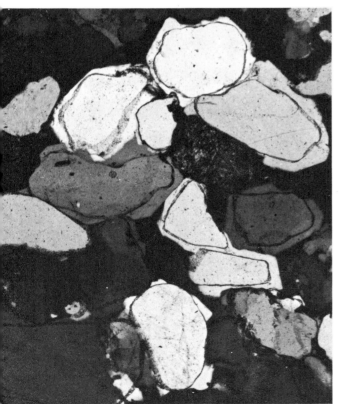

LEFT Sandstone thin section This thin section shows just how packed with quartz grains a sandstone may be. The dark areas between the grains are the cementing materials. Sections like this are photographed to identify specimens.

ABOVE Mudstone hand specimen Mudstone is a finely made-up sedimentary rock which usually occurs in massive non-layered form. As its name suggests, mudstone is formed when layers of silt and mud are subjected to great pressure over long periods of time.

BELOW Sandstone hand specimen Sandstone is composed of small quartz grains which have been cemented together. In this specimen the individual quartz grains can just be distinguished. There are a large number of different sandstones.

derive much information from his specimens of sedimentary rock. He should be able, by careful examination, to deduce the conditions under which the specimens were formed. Fossils are an important feature of sedimentary rocks, and these too provide a wealth of evidence about the plant and animal life prevailing at the time of formation. (There are a number of excellent reference books available to the amateur geologist, which will assist in fossil identification.) The map on page 73 shows the location of the sedimentary rock areas of the world.

The accumulation of debris

Rocks exposed at the earth's surface are constantly worn down by a variety of processes. These can be divided into two main groups – the process of weathering and the process of erosion. The term weathering is applied to those processes which cause the exposed rock to break down into smaller pieces or which cause the rock to dissolve and decompose. The fragment-producing process is usually known as mechanical or physical weathering, the decomposition process is known as chemical weathering.

During cold weather, especially at night in the winter months, the water present in the cracks, fissures and spaces in the rock will freeze. On freezing water expands by 9% of its volume. The ice formed applies a terrific pressure (approximately 2000 lbs per square inch) within the rock. As the temperature rises the ice melts and the pressure is released. More water enters the slightly enlarged space, the freezing process is repeated and pressure is again applied within the rock spaces. If the process is repeated often enough, fragments of the rocks are broken off. The freezing of water in household plumbing systems is a similar process and, all too frequently, yields similar results! The water expands on freezing and fractures the pipe in so doing.

This freezing-thawing cycle produces the steep, inclined accumulation of broken, irregular rock fragments which are known as scree and which are found in most mountain areas where the climate is cold enough for ice to form. Screes also occur in warmer areas, where they accumulate in a similar manner. The rocks are heated by the sun during the day and rapidly cooled at night or by sudden rain storms. The heating and cooling processes cause the rocks to alternately expand and contract. Pressures created by the expansion and contraction result in the splitting of fragments from the rock. Farmers used to use this method to get rid of large unwanted boulders. A fire would be lit around and under the boulder. When it was very hot, cold water was poured over it. This caused the rock to split. The smaller fragments could be easily removed. Forest fires in areas such as California, Canada and Australia can also cause the fragmentation of rocks. Scree debris accumulated in these various ways is the raw material for the sedimentary rock known as a breccia. Plants and animals also aid the mechanical disintegration of rocks, although only on a small scale. The wedging effect of roots tends to prise the rocks apart, and the burrowing and digging activities of small animals can also be instrumental in bringing about fragmentation.

Chemical weathering is the process by which rocks are dissolved and decomposed. Water flowing over or through rocks may dissolve certain minerals in the rock, or may change their form. The dissolving or changing of the minerals within the rock weakens its structure and leads to fragmentation. The water which chemically attacks the rocks is usually acid in composition, and this speeds up the weathering process. Rain water becomes slightly acid when carbon dioxide

BELOW **Fish River Canyon, South West Africa** A sequence of sedimentary rocks piled up on each other. Some of the rocks are more resistant to erosion, and stand out as ledges. A scree blanket covers some of the outcrops. This is a typical environment for the production of breccia.

BELOW RIGHT **Clints and grikes above Malham Cove, Yorkshire** The clints and grikes were formed by the chemical weathering of limestone. This area lies above Malham Cove, a huge roughly semi-circular limestone cliff, over which a waterfall higher than Niagara once poured.

ABOVE **Hastings, Sussex**
A sequence of horizontally bedded sedimentary rocks, with coarse pebbles at the base of the cliff. This is a typical environment for the formation of conglomerate.

RIGHT **Lavernock, Glamorgan** A sequence of alternately bedded shale and limestone which has undergone substantial marine erosion. The shales have been eroded, and the limestone stands out as hard bands. The pebbles in the foreground have become almost spherical, due to the tides.

or sulphurous gases in the air are dissolved in it, and decaying vegetation also adds acid to it. Spectacular evidence of this kind of weathering process can often be seen in limestone country. The limestone 'pavements' above Malham Cove in Yorkshire consist of giant blocks and fissures (known as clints and grykes) and provide a good example of the effect of acid water on this kind of rock. Limestone is mainly composed of the mineral *calcite*. Its reaction to acid water is not surprising since one of the diagnostic tests of this mineral is that it effervesces and dissolves in weak acids.

The fact that rocks are dissolved in this way is demonstrated by the amount of dissolved material present in the water of rivers, streams, lakes and tap water. People who live in areas where the tap water is obtained from limestone or chalk areas are familiar

RIGHT **Pure limestone hand specimen** A limestone such as this which contains little shelly debris or other mineral matter is called a pure limestone. It is white in colour, and is composed of almost 100% calcium carbonate precipitated from sea water.

BELOW **Sole marks, Rhinns of Galloway, Scotland** Small sole marks are often seen on the underside of a sedimentary bedding plane, here a greywacke. The shape of these marks indicates that the flow of water bringing sediment into the area was from left to right. By collating all the information of this type, it is possible to reconstruct the environmental conditions of the time of deposition.

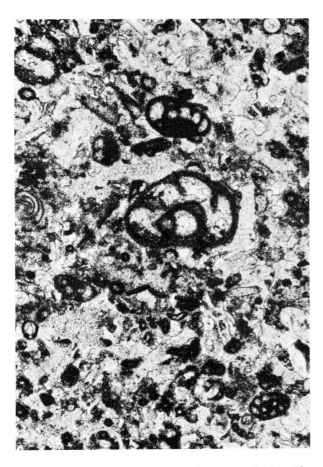

LEFT **Formaniferal lime-stone thin section** Limestone may be composed of myriads of tiny shelled animals (as are seen here). These are called formanifera, hence the name of such shelly limestone.

agent of all, since it dominates erosion in most areas of the world – even deserts sometimes have rain and glaciers do undergo some melting in the summer time. Torrential rain storms may sweep scree material down the sides of mountains, the irregular fragments becoming rounded by constant bumping against one another. These fragments collect at the base of the mountain as piles of rounded boulders known as piedmont fans. (The typical fan shape is sometimes seen in piedmont glaciers where one or more glaciers sprawl onto adjacent lowlands in an apron of ice.) If such a rounded pebble bed is buried, then the sedimentary rock known as a **conglomerate** is formed, although conglomerates can also be produced in other ways. If the scree material is swept into and along river valleys, the fragments are broken down into smaller and smaller pieces. This means that much finer sediment deposits are to be found in river valleys, these deposits often consisting of sand grains, which are highly resistant to wear and to chemical attack. Such sand deposits could form the sedimentary rock sandstone, but examples are rare. Sandstones are more commonly formed from marine or desert sands.

Great quantities of sediment carried by rivers are laid down at the point where the river enters the sea. Estuaries and deltas are areas of consolidated sediment build-up, for the quartz grains and the clay minerals are dropped by the river as it slows down to

with calcite deposits in water pipes and 'fur' inside kettles. The fact that weak acids also change the composition of some minerals is more difficult to demonstrate, but certain minerals – mica, feldspar, hornblende and augite for example – are changed into clay minerals by acidic waters. Because chemical weathering occurs in many different forms, its products are highly variable.

Weathering fragments and changes rocks but before the new material can be used in sedimentary rock building, it must be transported and deposited elsewhere. The removal of material from its place of origin is brought about by a number of processes known collectively as erosion. The individual processes are known as erosion agents – running water (rivers and streams), slow-moving glaciers, sea and wind are all examples of erosion agents. Erosion has a dual role. It transports weathered material away from the original site to a place of deposition, and at the same time assists the continuous processes of mechanical and chemical weathering of the fragments. The fragments, in their passage along rivers, locked in glacial ice, in the waves or in their wind-borne flight through the air, may strike and break other rocks over which they are passing. Thus it is often very difficult to draw a strict dividing line between products of weathering and products of erosion when studying the history of a particular sedimentary rock.

Climate is an important factor in both the weathering and erosion processes. In the colder regions of the world, ice is the important erosional agent. In these areas too, the freezing-thawing-freezing weathering process occurs. Wind is the most powerful erosional agent in desert and arid regions. The heating-cooling weathering process also occurs in these areas. Running water is probably the most important erosional

enter the sea. Over a period of time large shallow deltas are formed. The Nile delta, the Mississippi delta, the delta of the Ganges are good examples. Such sediment deposits can cause problems and it is not unusual to see boats dredging sediment from estuaries, so that ships can enter and leave the river in safety. Bristol is an excellent example of a port which has sediment problems, but most countries have harbours that are now defunct because of excessive silting. Material *not* deposited at the mouths of rivers may be

ABOVE **Coral limestone hand specimen** This limestone contains so many coral fossils that it is known as a coral limestone. It may have once been part of a coral reef.

carried out to sea and laid down on the ocean floor.

The delta, estuary and coastal areas that are occupied by vast expanses of forest – the tropical rain forest of South America, the Everglades of Florida in the U.S.A. and the mangrove swamps of southern Asia for example – are ideal sites for plant debris to accumulate. Providing the vegetation debris is rapidly covered, the sedimentary rock **coal** may be formed. All the continents contain coal deposits, deposits which were once vast forested seashore areas. European and North American coal was mainly laid down from tree accumulates during the Carboniferous period, about 300 million years ago. A study of the palaeogeographic map on page 13 suggests that both these coal areas were once part of the same tropical forest.

The sea is also an erosional agent, especially along the coast, where the waves beat continuously and relentlessly against the rock cliffs. Pebble or sandy beaches and mud flats are sediment accumulates formed by wave erosion, and the possible foundations of conglomerate, sandstone or mudstone rocks. The oceans, however, are more important as areas of sediment accumulation. Fragmental debris, derived from the eroding action of the waves or from rivers entering the sea, is laid down on the ocean floor. The finer debris material, such as clay particles, is deposited on the deeper ocean floors. Many of the shales and mudstone rocks found exposed in different areas of the earth's surface were formed from ocean floor debris. Huge earth movements uplifted the ocean floor and developed mountainous areas. As the centuries passed the mountains themselves were worn down by weathering and erosion, and the process was repeated.

The oceans are also areas where animal shells and skeletons may accumulate in thick deposits. Mussel and cockle banks and coral reefs are two good examples of of present day animal skeleton accumulations. It is easy to see how these shells and corals have become compacted together to form shelly and coral limestone sedimentary rocks. Shelly limestones, formed from the accumulation of vast numbers of shells, include brachiopod limestones, ammonite limestones and gastropod limestones. The Great Barrier Reef of the north east coast of Australia is probably the best known example of a present day coral reef.

At the present time, the upper levels of the oceans are full of microscopically small shells of both animals and plants. When the animals and plants die, these shells sink to the ocean floor and in areas where little other sediment debris is accumulating, they form thick deposits of the shelly sediments known as *oozes*. Some limestones of the past were formed under identical conditions to present day oozes, although the composition of those rocks can be seen only with the aid of a hand lens or a microscope. Another important group of oozes is also formed on the ocean floor.

ABOVE **Coral specimen** This specimen shows the thick calcite skeleton left by the zooid.

LEFT **Earth from space** The Nile valley dominates the foreground; deltas such as this are often found where sediment-laden rivers enter the sea. The Red Sea and the Sinai Desert can be clearly seen.

RIGHT **Great Barrier Reef, Australia** Animals which live together in a reef mass on the sea floor and which secrete a hard calcite skeleton belong to the coral group of animals. The small animal, known as a zooid, lives inside the calcite cup.

FAR RIGHT **Moon Valley, La Paz, Bolivia** Earth pillars have been carved out of soft rock by weathering and erosion processes. Compare these pillars with those shown in the photograph on page 15.

These are the mineral oozes which are formed from the minerals dissolved out of rocks by the processes of chemical weathering. The dissolved minerals are carried into the oceans and precipitated out in areas where the water contains a great amount of similar dissolved mineral material. The minerals then sink to the ocean floor and form a mineral ooze accumulate.

The Dead Sea in the Middle East is a good example of a mineral ooze. Its water contains so much dissolved mineral material that people may lie on the surface without fear of sinking. Although the Dead Sea provides an extreme example, it demonstrates the fact that dissolved mineral matter can become sufficiently concentrated to allow precipitation, and the floor of the Dead Sea has thick mineral ooze accumulates. Minerals which become dissolved in water and form mineral oozes are halite (salt), calcite, gypsum and iron. Rock salt, limestone, gypsum rock and ironstone are the sedimentary rocks which have their origins in such

precipitation processes. Evaporation of sea water can also result in precipitation of these mineral accumulates. For example, the famous salt and gypsum deposits of Stassfurt in Germany were formed when part of the sea became isolated as a lagoon and then evaporated. The rock salt deposits in Cheshire were also formed by the evaporation of a lagoon. Ironstone and limestone quarries provide examples of mineral-accumulated sedimentary rocks which are worked on a commercial basis. The hot springs, geysers and boiling mud pools which are found associated with areas of igneous activity contain a large amount of dissolved mineral matter. The Mammoth Hot Springs in Yellowstone Park in the U.S.A., among others, illustrate how readily minerals can be precipitated from such a hot water source (see cover photograph). The mineral in these springs is sulphur and it is building up large terraces of solid sulphur rock.

Dissolved mineral substances are not only present in the seas, in rivers and in volcanically-heated water but are also present in water which has seeped below ground level. This sub-surface water plays an important role in precipitating minerals between the pore spaces of sediment fragments and helps to cement these particles together. Calcite, silica and iron act as the main cementing agents in most sedimentary rocks.

Ice is another great debris collection agent. In the Carboniferous period, a mere 300 million years ago, glacial erosion and deposition occurred in South Africa, South America, India and Antarctica. At that time these continents were joined together near the South Pole. A glacier slowly grinds its way forward, gouging deep, straight-sided valleys, scratching, grooving and polishing the rocks along the valley floor as it moves. In areas under ice during the Great Ice Age, such features are typical of the once glaciated landscape. In older glaciated regions – the Sahara and the South Africa, South America, India and Antarctica complex, for example, deep scratch marks are the only remaining evidence of glacial erosion; the deep, straight-sided valleys have been redesigned by other erosional agents during the long intervening period of time.

As the glacier moves, it carries huge quantities of debris material along with it. At the snout of the glacier, the ice waste and debris accumulate into a feature known as a *moraine*. This marks the farthest limit of the glacier's travel. The sediment debris is called *till*, or boulder clay. Boulder clay is an apt name for such debris, for it consists of an accumulation

of both large boulders and extremely fine clay material. Boulder clay deposits left in valleys are not in the most favourable position for burial and transformation into sedimentary rock, but occasionally such deposits have been preserved in the geological record as till conglomerates. The evidence for the Scotland-Finland glaciation is based on the presence of till conglomerates only, and similar rocks are also found in the South Africa-Antarctica complex.

Wind, last but not least of the erosion agents, is more important as a sediment-depositing agent than as a wearing-away agent. The greatest effects of wind erosion are seen in areas where water and vegetation are sparse – the deserts and the areas surrounding them. In its wearing-away capacity, wind loaded with quartz sand grains acts like the great sand-blasting machines which are used to clean buildings in smoke polluted cities. These machines bombard the natural stone buildings with sand grains, thus removing the dirty outer layer. In a similar way, wind loaded naturally with sand grain sculpts and shapes the prominent rocks exposed above the general level of the desert area. Monument Valley on the Arizona-Utah border provides good examples of wind-abraded rocks. When the wind loses strength, the sand grains it carried are deposited in thick blankets and form 'sand seas'. Sedimentary rocks formed from such deposits are known as desert sandstones and are common rocks of the Permian and Triassic periods in Europe and North America.

ABOVE **Gypsum (Selenite) crystal** White and translucent, often with some play of colour, such minerals are found in some of the clay-type sedimentary rocks.

LEFT **Barland Quarry, Kittle, South Wales** Although beds are generally laid down horizontally, earth movements may uptilt the bedding planes to give dipping rocks. Note the variation in thickness between the layers.

ABOVE **Monument valley, Arizona-Utah border** This scene is typical of the effects of wind erosion in arid areas. The rock pillars known as 'totem poles' and the scree-blankets at their bases were carved out over centuries by the wind. Such scenes are well known to any fan of Western films.

Sediment-accumulated sedimentary rocks

Breccia is generally formed by the cementation of scree material, derived from the physical weathering of pre-existing rocks. It is mainly composed of large angular fragments, its composition depending on the nature of the pre-existing rocks, which can be igneous, metamorphic or sedimentary. The rock fragments are cemented together by minerals such as calcite, silica and iron, which are precipitated from ground water. Some breccia consists entirely of volcanic fragments and is known as **agglomerate**.

Conglomerate consists of rounded large rock fragments which are cemented together by particles of smaller size. Again, the composition depends on the nature of the pre-existing weathered rocks. The rounding of conglomerate material takes place as it is transported by running water. Conglomerate material may derive from pebble beaches, alluvial fans or glacial till deposits.

Sandstones often reveal interesting features regarding their origins and transportational histories, and there are many varieties present in the earth's crust. Generally, sandstones can be sub-classified into two main groups – pure sandstones and impure sandstones. Pure sandstones mainly consist of cemented quartz grains, and were usually formed in deserts or on beaches. Impure sandstones contain quantities of

RIGHT **Athabasca Glacier, Columbia Ice Field, Alberta, Canada** In cold regions with a heavy snow-fall, the climate sometimes permits the accumulation of extensive, almost mo-tionless snowfields. The light, powdery snowflakes melt, and refreeze into pellets of ice. New snow falls, increasing the pres-sure on the ice pellets. The grains recrystallize and weld together into solid glacial ice.

other minerals, and were usually formed as marine deposits. The sand grains are cemented together by various other minerals, and this cement material often gives the sandstone its particular characteristics. For example, if iron is the cementing agent, the rock has a red or brown colouration; glauconite is the cementing agent of greensand; other sandstone contains clay minerals; and quartzite is bonded by a silica cement. Shales, mudstones and siltstones are composed of very fine sedimentary material, and are usually dominated by clay minerals. Siltstones are usually lighter in colour and composed of slightly more coarse material than the others. Shales can be split into thin sheets, while mudstones are generally massive. The colouring of these rocks varies according to the impurities present in them. Small amounts of iron colour the rocks red; glauconite colours them green; manganese gives a purple colour; and plant remains give a black colour. Crystals of calcite, gypsum and pyrite are often present in rocks belonging to this group. Shale is the most abundant of all the sedimentary rocks, mudstones and siltstones are less commonly found.

Plant-accumulated sedimentary rocks

Coal and lignite are the most familiar plant accumulated sedimentary rocks, both of which were formed from forest debris buried in deltas and estuaries.

Animal-accumulated sedimentary rocks

Rocks belonging to this group are composed of a wide variety of animal shells and skeletons.
Shelly limestones are mainly composed of the shells of brachiopods, bivalves and gastropods, the shells of a number of microscopic creatures, together with precipitated calcite.
Coral limestones are composed of coral skeletons and crinoidal limestones are composed of crinoid stems.
Bone rock is formed from a vast accumulation of bones, scales and teeth of fish and reptiles.

Precipitated sedimentary rocks

Limestone consists almost entirely of calcite precipitated from ground water. Limestones are usually white in colour, but they can be discoloured – sometimes

ABOVE **Fold and fault, Denmark** Here the rocks have been folded into a nose-like formation, and the tremendous pressures to which they were subjected were so intense that faulting has also occurred. Two, if not more, faults are shown by the disruption of the brown, iron-stained bed. Layers of chalk, seen in the form of waves with black stripes of ash, stem from the Ice Age.

71

almost black – due to the presence of impurities in them. Chalk is a soft white variety of limestone.

Ironstone generally contains enough iron to enable the rock to be worked commercially and the metal extracted. Iron, precipitated from seawater, gives the stone its characteristic brownish-red colour. Fine clay minerals or sand grains are often the principle ingredient of ironstone. Ironstone may also occur as an oolitic mass of rounded iron grains set in a clayey mixture – the Cleveland ironstone found in Yorkshire and the ironstones of Newfoundland are examples. Rock salt and gypsum are formed by precipitation from seawater. These rocks naturally have the same properties and characteristics as the minerals of the same name, and are often found in association with each other.

Chert and flint are similar forms of impure silica, probably formed by precipitation, although many geologists are unsure of the method of formation. When fractured, flint splits with a splinter. Both flint and chert vary in colour from light grey to black and are often found as thin bands or as pebbles in limestone deposits.

Recognition of sedimentary rocks

In the process of sedimentary rock formation, layers of sediment are built up, layer by layer. Layering is therefore a fundamental characteristic of sedimentary rocks. Each layer, or bed, is separated from the

LEFT **Fault, St Agnes, Cornwall** Often the rock sequence is broken and moved so that the pieces on either side of the rupture line, the fault, no longer match. These faults are in Devonian grits and shales.

RIGHT **Map showing the distribution of sedimentary rocks throughout the world**

BELOW **Folding, Pyrenees, Spain** Limestones and shales folded into an unusual shape by the enormous pressures responsible for creating the Pyrenees.

ABOVE **Cliffs, Lulworth, Dorset** This highly folded sequence emphasizes the tremendous pressures which are involved in producing sedimentary rocks from sediments. The thicknesses of the bedding planes is seen clearly.

one above and the one below by a line of demarcation known as a bedding plane. This generally represents a sudden change in the grain size or in the composition of the sediment being laid down. A sedimentary rock sequence may, for example, show a bed of limestone, a bed of shale and a bed of sandstone, piled one above the other and separated by bedding planes. Bedding planes are also evident in layers of rock of the same material, suggesting that although the same sediment

was deposited, some change occurred in the depositional pattern. Variation in the thickness of the bedding planes is another feature of sedimentary layering. Very thin layers are called laminae. The thicker layers are called beds. The different layers may show variations in texture or colour which may be the result of seasonal climatic changes or fluctuations of sea level. As long as the bedding planes remain undisturbed, the last-formed stratum will always be found at the top, and the first-formed will be located at the bottom.

Sediment was usually deposited on a horizontal surface, and so horizontal bedding planes are commonly found. However, earth movements have often tilted the rocks to give inclined beds. In some areas earth movements have twisted and contorted the rock to give folded bedding planes.

Sedimentary bedding planes often show ripple marks, sun cracks or rain pit marks, all of which were formed before the rock became consolidated, and which give clues to the areas in which the rocks were formed.

Sedimentary rocks are often more colourful than igneous or metamorphic rocks. Sometimes this colour range is the result of the wide variety of rocks from which the weathered fragments come. Sometimes it is due to the cementing material that fills the space between the grains. Iron, for example, may stain the rock red, brown, pink, yellow, purple or green. Organic matter may colour the rocks from light grey to black.

A unique feature of sedimentary rocks is that they contain fossils – the preserved remains of animal and plant life. Fossils are an important indication of the age of the rocks, and also yield valuable evidence about about the animal and plant life present on earth during prehistoric times.

Chapter 4
FOSSILS

The word 'fossils' is a general term and is used to indicate a wide variety of animal and plant preserved material. Fossils may be the original hard parts of organisms, some as old as 100 million years. A bone or a seashell preserved from the distant past is a fossil. Logs of wood from prehistoric forests are also fossils.

The soft parts of organisms tend to decay very quickly after death, and so complete specimens of prehistoric creatures are seldom discovered. Fascinating exceptions are the complete bodies of Ice Age mammoths and rhinoceroses, preserved, complete with flesh, in the frozen tundra of Siberia and northern Canada. These too are fossils.

Fossils are also petrified (turned to stone) hard parts of organisms. Petrification occurs in different ways. For example, the pores of bones or shells may be invaded by percolating ground waters which deposit minerals within the organic material. Bones of dinosaurs have been preserved by this permineralization process.

In some cases, mineral matter from ground water has actually replaced the original organic substance. This process, known as replacement, is responsible for petrified wood fossils. Sometimes petrification involves the loss of an organism's volatile elements and the concentration of its carbon content, the residue forming a perfect outline of the original. This process is called carbonization, and there are many examples of beautifully etched fossil leaves formed in this way.

Tracks, footprints and burrows are also fossils. In the shaly sandstone of the Connecticut River valley in the United States there are thousands of dinosaur footprints, and worm burrows and casts are among the oldest fossils known.

Sometimes an organism beds into a rock, but later disappears, leaving an empty space which is the exact replica of the original organism. Such moulds are often found in Baltic amber. A natural cast, formed when such a mould was subsequently filled with mineral material, is also a fossil.

Fossils may be preserved in marine and fresh water sediments, in Arctic tundra, in lava and volcanic ash and in many other burial places. Although there are many forms of life, which strictly speaking, are neither plant nor animal, nevertheless, these two kingdoms serve as the classification under which most fossils are grouped.

Fossils – the preserved relics of past life

The study of the animal and plant life of the geological past not only forms the basis to a fascinating hobby,

Ammonites This specimen is of limestone rock containing a number of well-preserved examples of the now-extinct sea creatures known as ammonites. The nearest relatives to the ammonites living in the present day are the squid and the octopus. Ammonites were abundant in the seas and oceans during the Mesozoic Era.

but enables the geologist to subdivide geological time. Most of the major groups of animals and plants living today had relatives living millions of years ago – some as many as 600 million years ago during the Cambrian period; some groups, such as the algae and bacteria, have a geological record (although a rather sparse one) which began some 3000 million years ago. Throughout geological time the various animals and plants have undergone changes and by recognizing and analyzing these changes, the palaeontologist can subdivide geological time up into more easily discernible units. For instance, the largest group of present day living creatures, the arthropods, represented today by. the insects, crabs and millipedes, were represented in the Lower Palaeozoic Era by trilobites and scorpion-like forms called eurypterids. Ferns and horsetails were characteristic plants of the later Upper Palaeozoic Era, although they were rather different from present day ferns and horsetails. In the Mesozoic Era

reptiles, in the form of dinosaurs, were the dominant land creatures. Although there are many present day examples of reptiles, the dinosaurs themselves became extinct at the end of the Cretaceous period some 100 million years ago. No one really knows why, although suggestions have included lack of food, inability to breed and rear young, inundation by water and inability to adapt to dramatic climatic changes. So important is fossil dating to geological history that general terms such as the Age of Trilobites, the Age of Ferns, the Age of Dinosaurs are acceptable methods of representing periods such as the Lower Palaeozoic, Upper Palaeozoic and Mesozoic.

Thus fossils have been a useful aid in the delineating of geological time, but fossils can also give the geologist a more complete and interesting picture of the animal and plant life found on earth during those times.

Unfortunately not all animals and plants are easily preserved in the fossil record. Animals without shells, or skeletons – the soft-bodied forms such as jellyfish, worms, and so on – are rarely preserved. In these examples, the soft parts are totally destroyed by bacterial action as they are seldom buried quickly enough for preservation to take place. Since bacterial decay of shells and skeletons takes a longer period of time, the chances of preservation are rather better. It is likely that the fossils so far found represent only a fraction of the actual life which existed in prehistoric times since the environment in which the animal died

or in which it was trapped must determine its preservation or destruction.

How and where fossils are preserved

Although the soft parts of animals and plants are not usually preserved in the fossil record, there are a few relatively recent and certainly spectacular examples of this kind of preservation. The Woolly Mammoths that roamed over most of Northern Europe about a million years ago in rather colder conditions than at present have been preserved in the frozen soils of Siberia. The bodies of these animals have been literally deep-frozen and modern hunters have reported that their sleigh dogs have eaten, with evident relish, the million-year-old meat! California is another important fossil locality. Here many different animals – antelopes, sabre tooth

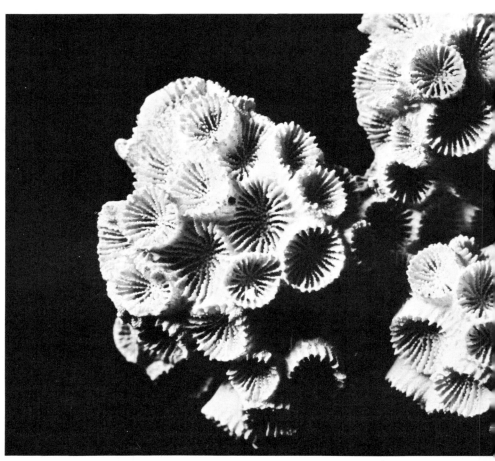

RIGHT **Gastropods** (*Turitella*) These gastropod shells are about 40 million years old. The creatures inhabiting the shells belonged to the same family as today's garden snails. Under certain circumstances, these fossil shells may be so abundant that a gastropod limestone is formed. Note the much smaller shells intermixed with the larger gastropod shells, and the varying sizes of the gastropod shells themselves. Note also how some shells have survived intact in fossilized form.

LEFT **Fossilized fish** These fish probably died about 200,000,000 years ago. The fact that so many died together is probably evidence of some major catastrophe – such as the drying out of their environment or the water suddenly becoming too hot or too cold, too fresh or too salty.

RIGHT **Echinoid** (*Plymosoma koenigi*) A beautifully preserved example of a sea urchin, showing a few of the heavy spines used for locomotion and protection.

ABOVE **Amber with embedded bee** (*apis mellifera*) Amber represents the fossilization of sap which has been released from a pine tree, and solidified. The unfortunate bee was trapped by the sticky sap, in which it was enveloped and preserved as the sap solidified. This manner of preservation is quite rare, but examples do exist. Similar methods of preservation are sometimes used to manufacture unusual pieces of 'jewellery'.

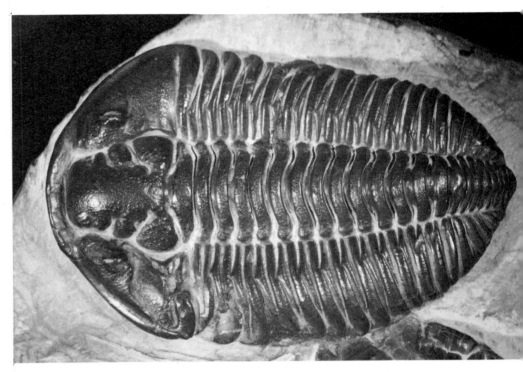

tigers, elephants, and even birds – are well preserved
in the La Brea Tar Sands. This tar swamp entrapped
any animal that strayed into its sticky quagmire
(probably attracting other creatures, innocently seek-
ing an easy meal). The creatures slowly sank into the
swamp; the soft parts – the flesh of the animals and the
feathers of the birds – being protected from decay by
the slightly acid nature of the tar. A further example
of the preservation of soft parts can be found in
examples of insects entombed in amber. Conifer trees
often weep sap from their bark. Insects, such as flies,
bees or ants, can become trapped by this sap, and
quickly sealed off from outside bacterial decay. When
the sap finally solidifies into amber, completely preserved insects will remain entombed.

Hard skeletal material such as shell or bone is
preserved by one of two major methods. The first involves the replacement of the original hard packed
substance by a new material. For example, the calcite
shells of brachiopods, the calcite coils of ammonites,
and the calcite cases of sea urchins are often replaced
by quartz and iron pyrite minerals. This process results
in the formation of very hard quartz shells or 'fool's
gold' skeletons. This fossilization process takes place
when the shells are deeply buried in sediment. Per-

colating mineral-rich waters replace calcite of the
original shell for silica or pyrite. The process is not
restricted to originally calcite shells. The protein coat
of trilobites for example can be replaced by pyrite –
giving a beautiful golden-coloured trilobite, a fossil
well worth discovering. If the process of replacement
does not occur, then the shell material will be recrystallized and strengthened by more of the original
material – again probably derived from percolating
waters at depth.

Graptolites, an extinct group of animals that lived
during the Lower Palaeozoic Era, are often found as
thin pencil-like markings on fine grained rocks of this
period. Graptolites have no known living relatives at
the present time, and their method of preservation is
called carbonization. The original skeleton is composed
of oxygen, hydrogen, nitrogen and carbon, but all the
contents except the carbon are lost. Plants are also
preserved in a similar manner for coal is another form
of carbon, formed from compressed layers of plant
debris.

In order to become fossils, plants and animals must
die in particular areas. They must be quickly covered
by sediment in order to stand the slightest chance of
preservation, and so areas of rapid sediment cover are
best for preservation. Deltas, river estuaries and the
sea, where sediment accumulates, are therefore the
best preservation areas. Sediments that are most
likely to be converted into sedimentary rocks are
again the most likely fossil areas. The three types of
area mentioned above are generally the most common
of the sedimentary rock environments. Areas of sediment on land – such as scree deposits, rivers and sand
dunes – are far less likely to be converted to sedimentary rock.

Generally speaking, it seems that the fossil record is
heavily biased in favour of those animals and plants
which had hard outer coverings, and is especially
favourable to those life forms which lived near or in
the sea. Those animals and plants with soft outer
coverings, and which lived inland are much less likely
to be preserved in the fossil record.

The collection and identification of fossils

Most sedimentary rocks from the Cambrian period to the present day will reveal some evidence of former life. Fossil hunting is similar to mineral hunting in the sense that some localities are more favourable than others. A certain amount of patience is required if good, well-preserved, undamaged specimens are to be found. Limestone and the shales are probably the best type of rock to begin a search for fossils. Sandstones and conglomerates are generally poor in fossils, since the areas in which these rocks were formed were unlikely environments for abundant plant or animal life. A good place to search is among the debris at the base of a quarry or similar exposure, since specimens are likely to have weathered completely or partially out of the rock in such localities. A small chisel and a geological hammer are very useful for extracting the fossil from the rock matrix. If the rock debris at the base of an outcrop is barren of fossils, it is usually an indication that the rocks themselves are poor in fossil material and that one should try elsewhere.

A visit to a local museum will often provide evidence of exact fossil locations, as fossil samples usually carry

ABOVE **Crinoid stems**
Crinoids lived in a cup-like structure at the end of a flexible stem. They are popularly referred to as 'sea lilies' from their fanciful resemblance to flowers.

RIGHT **Fossilized tree, Stanhope, Co. Durham**
This fossil dates from the Carboniferous period. The trunk and roots indicate that these fern-like trees reached heights of more than 100 ft.

a name tag describing their place of origin. A visit to a museum will give you an idea of the range and types of fossils to be found and will also save wasting time and energy in a fruitless search of poor areas.

A careful list of the fossils should be kept in a book and the fossil should be properly labelled or numbered. Each fossil should be carefully wrapped in tissue paper and marked with the locality from which it was obtained; if this is done then the relative age of the rocks from which it was collected can be found by comparing the collected fossil with one whose age is known.

Fossils obtained from Mesozoic or younger rocks are often easier to extract than fossils from older strata; but by the same token the fossils are often far more fragile and need extra care in their handling and displaying. In such cases the fossil should be carefully wrapped and if possible coated with nail varnish, shellac or sprayed with aerosol lacquer.

Fossil hunting can be an extremely rewarding and interesting hobby. The prehistoric record of plant and animal life provides a fascinating contrast to present life forms on earth, and helps us to unravel the secrets of their development.

RIGHT **Brachiopods**
Mesozoic brachiopods which have been naturally weathered out of the rock. Brachiopod fossils such as these are now very rare.

BELOW **Fossilized fern**
Impressions of fern leaves in a coal measure dating from Carboniferous times. These 'ferns' have been fossilized by the carbonization process.

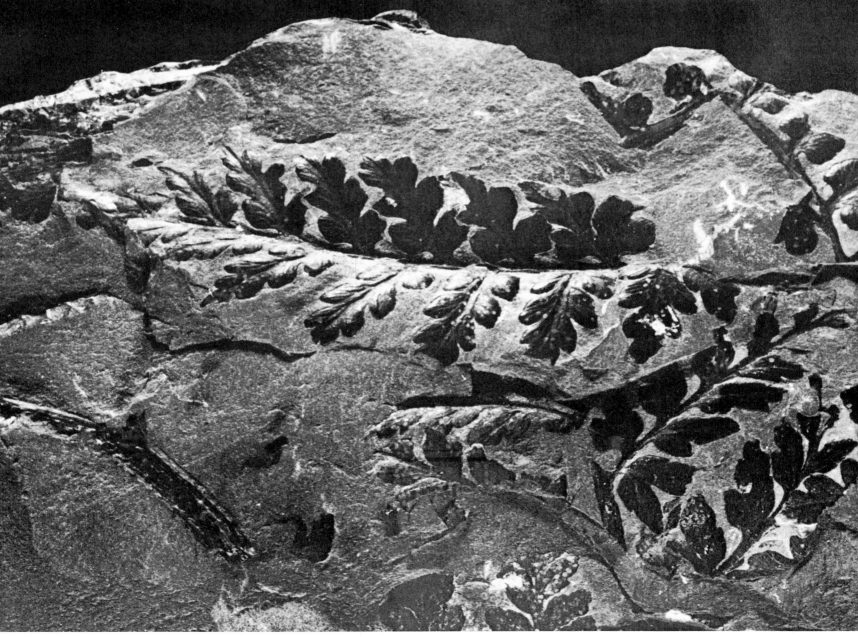

Chapter 5
METAMORPHIC ROCKS

The third and last major group of rocks present in the earth's outer layers are the metamorphic rocks. These are formed when rocks already in existence are subjected to intensive heating, fluid impregnation or pressure. Under extremes of heat and pressure, physical and chemical changes occur in pre-existing rocks. The process of change is known as metamorphism, and generally occurs deep within the earth's outer layers. Because of this most metamorphic rocks are only revealed after a considerable amount of erosion has taken place.

Both sedimentary and igneous rocks can be changed into new rock types by the processes of metamorphism. For example, the metamorphic rock, slate, is derived from sedimentary shales and mudstones. Sedimentary limestones and shelly limestones can be metamorphozed into marble, sedimentary sandstones into metaquartzite. Metamorphic rocks derived from pre-existing igneous rocks include schist, hornfels, cataclastite, gneiss (pronounced 'nice'), serpentinite and soapstone. Previously metamorphozed rocks can also be metamorphozed for a second time.

The extensive heating needed for metamorphism can be provided from a number of different sources. Liquid magma can intrude into the rocks in the earth's outer layers; movements of huge rock masses deep within the earth can provide quantities of frictional heat; heating may occur as a function of the depth at which rocks are buried. Permeating hot fluids and gases are often released during the heating and pressurizing processes, and it is these influences which are generally responsible for the formation of new minerals – garnet, kyanite, chlorite, staurolite, and talc, which are an important feature of metamorphic rocks.

The intensive pressures required for metamorphism are a function of the depth at which the pre-existing rocks are buried, or are derived from the effects of earth movements. It is these pressures which are responsible for the varying textures which are another important feature of metamorphic rocks. It is these varying textures which provide a simple basis for the classification of metamorphic rocks, and which help to simplify the problems of recognition by the amateur geologist.

Metamorphic rocks can thus be divided into two main groups – banded (or foliated) and non-banded (non-foliated). Banded rocks can be split into sub-parallel sheets. If the banding is frequent (approximately 6 mm or less), so that the rock can be split into thin, flat, smooth sheets along the bands, then the rock has the property known as slaty cleavage. This

South Stack, Anglesey, North Wales These rocks, well exposed in the steep cliff face, were originally sedimentary. They were converted into metamorphic rocks by heat and pressure. Some concept of the pressures involved can be gained from the intensely folded nature of the rocks.

RIGHT **Thermal metamorphism** The area affected by thermal metamorphism is generally relatively local. The heat from an intruding magma source alters existing sedimentary rocks. Metamorphism takes place to differing degrees, depending on the nearness of the metamorphozed rocks to the heat source.

CENTRE RIGHT **Cataclastic metamorphism** The area affected by cataclastic metamorphism is local over a narrow zone. Earth movements create tremendous local pressures.

FAR RIGHT **Regional metamorphism** The area affected is extensive, often many hundreds of miles square. Regional metamorphism happens when movements take place deep within the earth's crust.

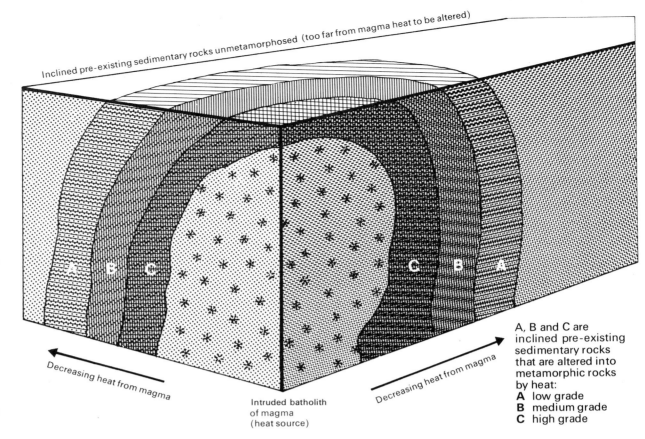

Inclined pre-existing sedimentary rocks unmetamorphosed (too far from magma heat to be altered)

A B C C B A

Decreasing heat from magma

Decreasing heat from magma

Intruded batholith of magma (heat source)

A, B and C are inclined pre-existing sedimentary rocks that are altered into metamorphic rocks by heat:
A low grade
B medium grade
C high grade

ABOVE LEFT **Mica-schist hand specimen** Mica-schist is a rock composed almost entirely of the mineral mica. This is the mineral which is responsible for the glistening white appearance of the rock. The planes of schistose cleavage can be seen quite clearly on the left hand side of the specimen.

ABOVE RIGHT **Slate hand specimen** The fine-grained nature of this metamorphic rock can easily be recognized in this specimen. The cleavage, which is characteristic of slate, can be seen at the top of the specimen.

splitting, as the term implies, is a feature of slate, and is the main reason why this rock is such a useful material in the manufacture of roofing tiles. The narrow banding of slate is sometimes difficult to distinguish from the thin bedding planes that occur in some of the sedimentary rocks. Such sedimentary rocks are often used for roofing material, and are known to the quarrymen as 'slate' although they are in fact thinly bedded siltstones or limestones.

The metamorphic rocks are divided into two groups based on the presence or absence of foliation. The first group, the foliates, show three types of banding: the slate has perfect planar parting, the schist has undulose bands, the foliation sheets usually being covered with micaceous minerals, and the gneisses are also characterized by undulose undulating planes to be more closely welded so that there is no easy parting definable between them.

The second group, the non-banded metamorphic rocks, lacks banded textures. They are composed of either a mosaic of interlocking mineral grains (marble, metaquartzite, hornfels) or show some evidence of a shattering or a crushing of the minerals (cataclastite).

The type and character of metamorphic rocks depend on which of the two variables of heat and pressure dominates during their formation, and on the intensity of these variable influences. Three types of metamorphism can be recognized, depending on the dominant influence, and each of these three types can be sub-divided into low, medium or high grade, depending on the intensity of the influences involved. The three types are thermal metamorphism, cataclastic metamorphism and regional metamorphism.

Thermal metamorphism is that type of metamorphism in which heat from a hot magma intrusion is the dominant influence. Cataclastic metamorphism occurs

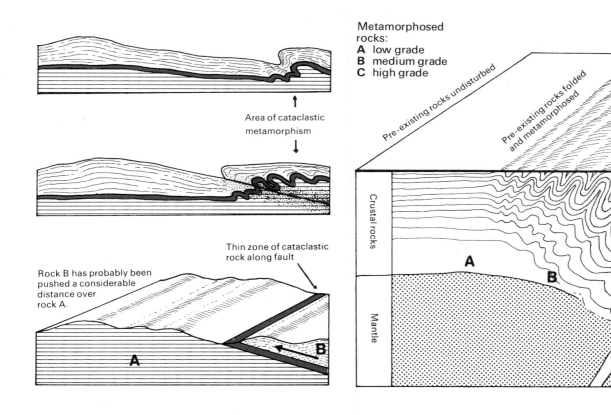

Metamorphosed
rocks:
A low grade
B medium grade
C high grade

Decreasing intensity

Area of cataclastic
metamorphism

Thin zone of cataclastic
rock along fault

Rock B has probably been
pushed a considerable
distance over
rock A.

Pre-existing rocks undisturbed

Pre-existing rocks folded
and metamorphosed

Ocean basin

Crustal rocks

Mantle

A

B

C

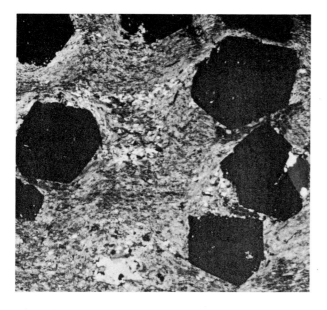

CLASSIFICATION OF METAMORPHIC ROCKS

Texture	Cleavage	Metamorphic rock type
Banded (foliated)	Slaty cleavage	Slate
	Schistose cleavage	Schist
	Gneissose cleavage	Gneiss
Non-banded (non-foliated)	Crushed minerals	Cataclastite
	Mosaic of minerals	Marble Metaquartzite Hornfels

when the dominant influence is the heat and pressure generated from the friction during earth movements. Regional metamorphism occurs when heat and pressure are functions of the depth at which the pre-existing rocks are buried. These three types of metamorphism are illustrated in the diagram above. The dominant influences of heat, pressure and permeating fluids can act in various combinations and with varying intensities, and so the conditions under which metamorphic rocks are formed vary tremendously. This means that both the type and the character of metamorphic rocks are also very variable.

The metamorphic rocks

Slate is formed from fine-grained sedimentary rock such as shale and mudstone. The small minerals present in these pre-existing rocks (such as mica) are rearranged under metamorphism into thin parallel layers, or bands. Any new minerals which are formed during the metamorphic processes (such as talc, chlorite) are similarly aligned. It is along these closely-spaced bands that the rock may split or cleave into thin smooth sheets. Slate is quarried in many parts of the world, and is an ideal building material. Its strength and the fact that it splits into thin sheets are obvious building qualities, but slate also *looks* attractive. The colouring varies, according to mineral content and metamorphic history. Blue, grey, green, red and black slates occur in different parts of the world. The photograph overleaf shows quarry workings in North Wales, but one quarry is very much like another and the scene is typical of other areas in Britain, where important slate quarries are found. The Lake District and the southern uplands of Scotland are two such areas. Important slate quarries are found at Willunga in South Australia and at Coulborn in New South

ABOVE LEFT **Garnet-schist thin section** A very thin section cut from the rock and examined and photo-graphed under a polarizing microscope. The large well-formed crystals of garnet appear as black straight-sided shapes of varying sizes. They are set in a fine-grained, wavy sur-round of other minerals, including quartz and muscovite. All these minerals are orientated in the same direction.

ABOVE **Classification of metamorphic rocks** Texture and cleavage are diagnostic.

Wales. In America, quarries are located in Maine, Pennsylvania, Vermont and Virginia. All the areas mentioned so far contain slate formed during the Lower Palaeozoic Era and which is between 400 and 600 million years old. More recent slates, formed about 50 million years ago in the Eocene period, are quarried at Chiavari in Italy, and in Switzerland.

Slates generally form as a consequence of low-grade regional metamorphic processes. Any fossils originally present in the sedimentary rocks are either destroyed or highly distorted by the intense heat and pressure occurring during metamorphism.

Schist is a medium-grained metamorphic rock and the minerals can be more easily seen than in slate. The foliation, or banding, is not as thin or as smooth as in slate. Schists are generally formed from shales as a result of medium-grade regional metamorphism, but they may also be formed from basic igneous rocks as a result of low-grade regional metamorphism. A glistening, mica-rich surface is a characteristic of schist, and thin sections often reveal both the schistose cleavage and the abundance of mica which is present. Other than mica, schists often yield good specimens of chlorite, and more particularly specimens of the gemstone garnet.

Gneiss is a common coarse-grained metamorphic rock consisting of alternate coloured bands. Gneiss resembles granite in appearance and in fact contains the same minerals as granite. The banded appearance of gneiss, however, provides the necessary clue to enable the amateur geologist to distinguish between the two rocks. The lighter coloured bands in gneiss contain the minerals quartz and feldspar, while the dark bands contain black mica (biotite) and hornblende. Gneisses are usually the result of high-grade metamorphism on sedimentary shales or on igneous rocks, but low-grade previously metamorphozed rocks can also form gneiss as they undergo a second metamorphism. Areas of highly-deformed rock are characteristic of gneiss, as the high-grade processes twist and bend the rock into weird shapes. The Scottish Highlands, Scandinavia, the Transvaal and the Catalina Mountains in Arizona are areas rich in gneiss.

Of the non-banded metamorphic rocks, **marble** is probably the most familiar. This beautiful material is deservedly popular with both sculptors and builders. The texture varies considerably. The famous Carrara marble from Tuscany in Italy has a white sugary texture, composed of a mosaic of interlocking calcite minerals. It originates from the metamorphism of pure sedimentary limestones and shelly limestones. Similar white marble occurs where the Whin Sill thermal has intruded carboniferous limestone, at Paros in Greece, Ruszikaer in Hungary, Peak in Malaya and in the Austrian Tyrol. The more common multi-coloured marbles, used in ornamental building work, are derived from limestones that contained impurities in their original forms. The light green marble found in Connemara in Ireland is world famous, as are the black marbles of Mayenne in France, Namur in Belgium and Kilkenny in Ireland. Barstow, Crestmore and Twin Lakes in California are other areas which contain a wide variety of coloured marbles.

Most marbles are formed by either thermal or regional metamorphism. Any fossils contained in the original limestone are converted into a granular

LEFT Slate quarry, Nantlle, North Wales A typical slate quarry. Lower Palaeozoic rocks have been pressured into slate. With its thinly cleaving plates, slate is an important roofing material.

BELOW Himalayas, India A panoramic view of this relatively young mountain system – less than 100,000,000 years old – in the north of India. Such mountain areas often show extensive suites of metamorphic rocks.

RIGHT Marble quarry, Carrara, North Italy One of the most famous marble quarries in the world. Huge slabs of pure white calcite result from the metamorphism of pure sedimentary limestone.

BELOW RIGHT Gneiss, Lugard's Falls, Kenya The thick alternating bands are typical of gneiss. The lighter bands are of quartz and feldspar minerals; in the darker bands hornblende and biotite minerals occur.

mosaic of calcite, and thus destroyed. Both marble and limestone react to dilute acids, and both have a low hardness grade. Differences in texture however enable the amateur geologist to distinguish between the two. More scientific differentiations are possible. Marbles are more coarsely crystalline in nature than limestones and lack fossils. The limestones, on the other hand, are fine-grained and fossil bearing.

Metaquartzite is rather similar to marble, in that it has a white sugary texture. However, metaquartzite, as its name implies, is composed of a mosaic of interlocking recrystallized quartz minerals, and this gives the rock its characteristic texture. Metaquartzites are very hard rocks, since they are composed mainly of quartz which has a hardness factor of 7 (see table on page 30), and have a smooth break pattern. Since the quartz grains are welded together, the rock tends to split through the grains. Metaquartzites can result from thermal and regional metamorphism of both pure sandstones and impure sandstones.

Hornfels is fine-grained, and is composed of a variety of tiny interlocking minerals in a mosaic pattern. Heat from magma intrusions on most rocks will result in the formation of hornfels, and the type of minerals present depends on the original rock. The most common

hornfels are developed from the baking of shales, so that they have a composition similar to slate, although they lack the slaty cleavage which is a slate characteristic. Such shale-hornfels characterize the Odenwald, Andlauvosages, and Hessen areas of Germany where the shales were metamorphozed by granite batholiths. Geologists think that the Donegal (Ireland) and Westland (New Zealand) hornfels were formed in the same way. The hornfels rock on Mull, Scotland and the Clearwater area of California may have been formed through those conditions prevailing during the formation of igneous dykes (see page 38).

Cataclastite is a metamorphic rock formed by the frictional forces which are present during cataclastic metamorphism. The common minerals contained in the pre-existing rocks are broken and crushed by the frictional action, and it is this crushing which is the major characteristic of cataclastite. Rocks belonging to this family are found in thin zones in highly-deformed faulted areas of rock, and are common in Scotland, North Wales, Devon and the Lake District of Britain.

Two closely related metamorphic rocks also occur. One contains quantities of the mineral serpentine, and because of this is known as **serpentinite**. The other is dominated by the mineral talc, and is known as **soapstone**. These rocks have the same characteristics as the dominant mineral, and both are developed from an olivine-rich igneous rock during low-grade regional metamorphism. They may be foliated or non-foliated, so they are rather difficult to place in the classification of metamorphic rocks on page 85.

Recognition of metamorphic rocks

The location of metamorphic rocks is an important clue to their recognition. The usual location sites for metamorphic rocks are around the larger igneous

intrusions, or over larger areas which have been highly deformed by earth movements. Most of the rocks labelled 'Pre-Cambrian' on geological maps are of metamorphic origin – another clue in the location of rocks belonging to this group. The major areas of metamorphic rock location are shown on the map opposite. The characteristic textures of metamorphic rocks – the banding or foliation, the crushed minerals or the mosaic patterns of interlocking materials – are further aids in their recognition. The presence of minerals such as garnet, talc, chlorite, kyanite and staurolite also provides evidence of metamorphic rocks, particularly if these minerals are well-formed and present in large quantities.

The three major groups of rocks – igneous, sedi-

TOP **Cataclastite hand specimen** Cataclastite is formed mainly by the crushing pressures involved in mountain-building; the original rock is shattered into fragments by the grinding action of two sections of the earth's crust. The fragments form cataclastite.

ABOVE **Hornfels hand specimen** Hornfels is a fine-grained rock, which is non-foliated. It is formed mainly as a result of heat.

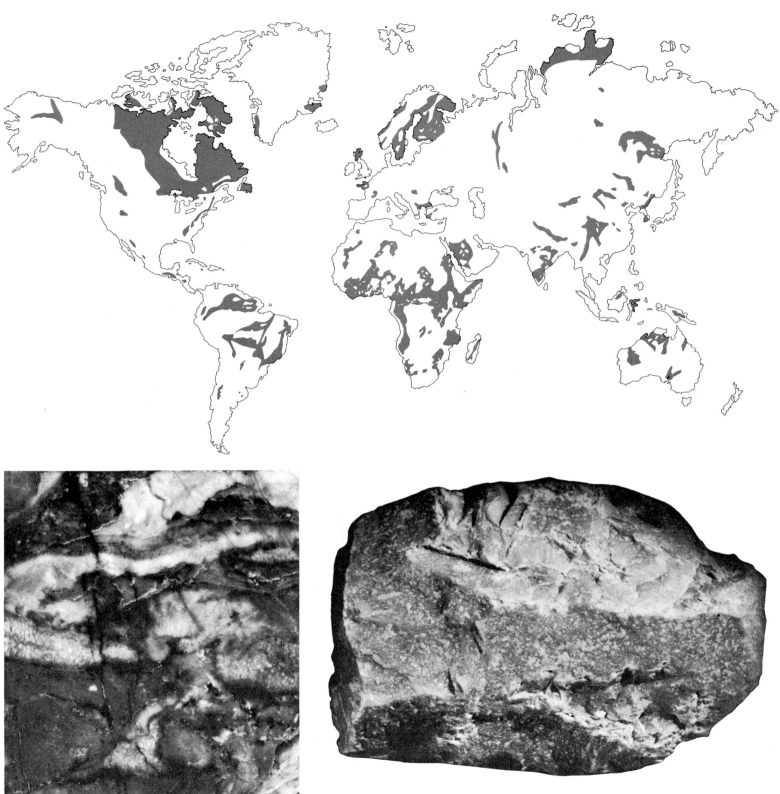

mentary and metamorphic – form an interesting study, and in most cases are easily collected and identified. The amateur geologist will have greater difficulty in collecting many of the specimens discussed in the next chapter, for they form that group of minerals which have fascinated and beguiled man for thousands of years – the minerals which are known collectively as gemstones.

If the mountain ranges of the world consisted entirely of diamond, emerald and ruby man would probably have found a way of building his houses and roads from these materials! But these, and other gemstones, are fascinating and valuable largely because of their rarity. It is this property which all the minerals discussed in the next chapter have in common.

TOP Map showing the distribution of metamorphic rocks throughout the world

ABOVE **Metaquartzite hand specimen** Metaquartzite is a metamorphozed form of quartz. The welded grains give some indication of the extremely hard nature of this rock.

LEFT **Marble hand specimen** A close-up of polished marble. In this specimen, the dominant mineral is calcite.

Chapter 6
GEMSTONES

Of the 2000 known minerals only a small number – about 80 – can rightly be called gemstones. In order to qualify for membership of the gemstone 'family', a mineral must possess three attributes not usually found together in one specimen. These are the attributes of beauty, durability and rarity. Many of the minerals in crystal form which do possess all three of these attributes – diamond, ruby, sapphire, emerald, opal – are familiar to us. Others, those minerals which possess qualities of beauty and durability, but are commonly found, are generally classed as semi-precious gemstones. Both precious and semi-precious gemstones are used to make beautiful jewellery. Of course, many minerals which are not classed in either category can be used to make attractive and decorative jewellery and artefacts. In fact, any interesting and well-marked stone can be used to make a ring, a pendant or a key-ring. Interesting lumps of mineral or small rock specimens can be made into paperweights, bookends, lamp bases and dozens of other attractive and useful objects.

'Beauty' is, of course, very much in the eye of the beholder. Different people see beauty in different things. But as far as gemstones are concerned, most people agree that their beauty lies either in the colour of the stones or in their light-refracting and reflecting qualities. For example, the pigeon blood colour of ruby, the grass-green of emerald, the cornflower-blue of sapphire are the colour qualities which make these stones beautiful. So important is colour that two minerals of the same chemical composition, but of different colours, may be classified as different stones: the red ruby and blue sapphire both have the same chemical composition – Al_2O_3.

On the other hand, one species of gemstone might be found in many different colours, due to small differences in the chemical composition: topaz may be yellow, blue, claret, brown or colourless; tourmaline may be black, blue, pink, green or red. The same colour may be found in many different gemstones: diamonds, beryl, corundum, topaz, quartz and orthoclase can all be yellow.

The beauty of a gemstone may lie not so much in its colour, but in its ability to refract and reflect light – diamond is one of the best examples of such a gemstone. Such beauty in gemstones is dependent on four optical phenomena – refraction, reflection, dispersion and pleochroism.

As a light ray passes across a surface of separation – the surface between two different media – into an optically denser medium, it is refracted, or 'bent'. This

Diamond The underground working at Kimberley, South Africa, where most of the world's diamonds are mined. This unusual view of the workings, showing a workman drilling into the rock, was taken through a diamond placed in front of the camera lens. Most of the diamonds mined are of industrial quality only.

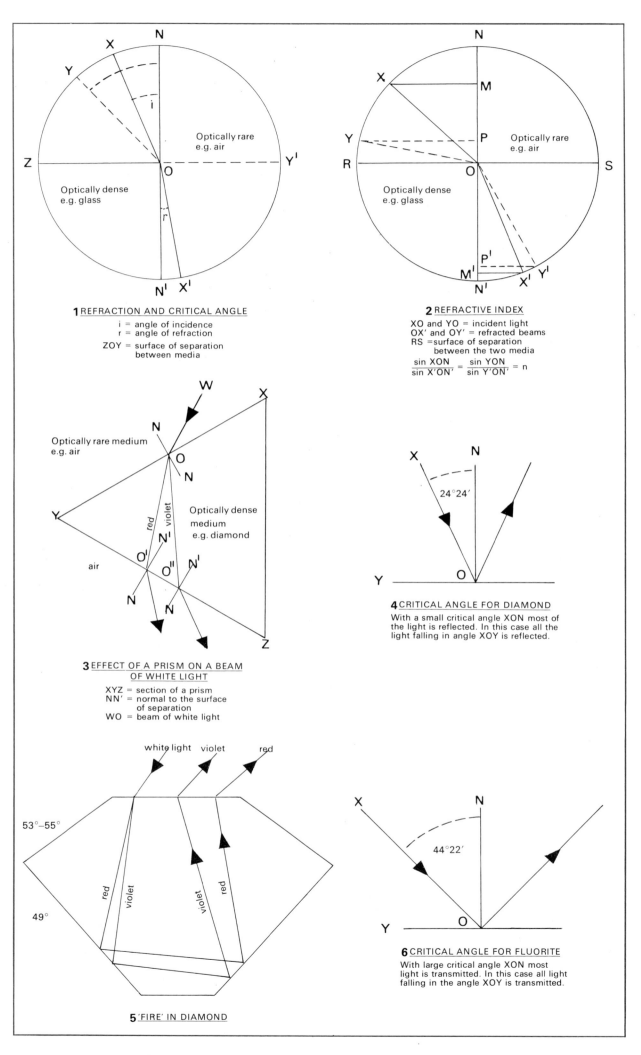

1 REFRACTION AND CRITICAL ANGLE

i = angle of incidence
r = angle of refraction
ZOY = surface of separation between media

Optically rare e.g. air
Optically dense e.g. glass

2 REFRACTIVE INDEX

XO and YO = incident light
OX' and OY' = refracted beams
RS = surface of separation between the two media

$$\frac{\sin XON}{\sin X'ON'} = \frac{\sin YON}{\sin Y'ON'} = n$$

Optically rare e.g. air
Optically dense e.g. glass

3 EFFECT OF A PRISM ON A BEAM OF WHITE LIGHT

XYZ = section of a prism
NN' = normal to the surface of separation
WO = beam of white light

Optically rare medium e.g. air
Optically dense medium e.g. diamond
air
red violet

4 CRITICAL ANGLE FOR DIAMOND

With a small critical angle XON most of the light is reflected. In this case all the light falling in angle XOY is reflected.

24°24′

5 'FIRE' IN DIAMOND

white light violet red
53°–55°
49°
red violet violet red

6 CRITICAL ANGLE FOR FLUORITE

With large critical angle XON most light is transmitted. In this case all light falling in the angle XOY is transmitted.

44°22′

7 Diamond — 24°24′

8 Zircon — 31°41′

9 Quartz — 40°8′

10 Diamond — 35° 41°

11 Zircon — 43° 40°

12 Quartz — 45° 41°

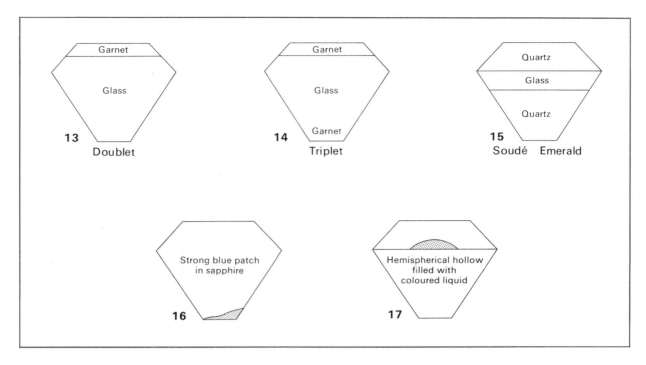

13 Doublet   **14** Triplet   **15** Soudé Emerald

16 Strong blue patch in sapphire

17 Hemispherical hollow filled with coloured liquid

FAR LEFT **1 Refraction and critical angle** Showing how light is 'bent' as it passes from an optically rare medium to an optically dense medium.
2 Refractive index.
3 Effect of a prism on a beam of white light The component colours are separated out by the prism.
4 Critical angle for diamond With a small critical angle most of the light is reflected; all the light falling within angle XOY is reflected.
5 Fire in diamond The diamond acts as a prism and separates the light into its component colours. This is called dispersion.
6 Critical angle for fluorite With a large critical angle most light is transmitted.

LEFT **The critical angles of three gemstones** – the smaller the critical angle, the more light is reflected.
7 Diamond. 8 Zircon.
9 Quartz.
Angles at which the gem must be cut to give maximum light reflection
10 Diamond. 11 Zircon.
12 Quartz.

RIGHT **Methods of simulating gemstones**
13 Doublet. 14 Triplet.
15 Soude Emerald. 16 The blue patch placed near the base of the pavillion gives total blue colour to the stone. 17 The coloured liquid in the hollow hemisphere colours the whole stone as light rays pass through it.

BOTTOM RIGHT **Below ground at Kimberley diamond mines** Diamond-bearing blue ground, blasted from its place in the volcanic pipe far below ground, is carried on a conveyor belt along an access tunnel to the shaft up which it is hoisted to the surface. This highly mechanical method of mining is much safer than earlier mining methods.

phenomenon can be observed by placing a pencil in a glass of water. The light passing through the surface of a diamond is also refracted. A constant ratio exists between the sine of the angle at which the light ray crosses the surface of separation and the sine of the angle of refraction. This is called the refractive index is a useful and important property. Except in those minerals belonging to the cubic system (see Chapter 1), the value of the refractive index varies with the direction of observation. When light rays cross the surface of separation between two media, passing from the dense to the rarer medium, there is an angle such that the refracted ray vibrates along the surface of separation. This is called the critical angle and it determines the amount of light which will be reflected from the interior facets of the gemstone.

When white light is passed through a prism it is split up into the spectrum colours of red, orange, yellow, green, blue, indigo and violet. These spectrum colours have different wave-lengths. The long waves – violet – are refracted more than the short waves – red – and so each colour reaches the eye fractionally later than the next, producing the familiar rainbow effect. This is the phenomenon of dispersion, and is responsible for the 'fire' which gemstones such as diamond, sphene and synthetic rutile show.

Pleochroism is the ability of coloured minerals to absorb light to different degrees depending on the direction in which the light is passing through the crystal. The importance of this property – colour change with direction – lies in the selection of the best direction of cutting to produce the best colour. For example, both ruby and tourmaline can occur as natural red crystals in an elongated hexagonal form. To obtain the best colour in ruby, the cut must be made parallel to the hexagonal end. But in tourmaline, the best colour is obtained from a cut parallel to the length of the crystal. Some gems – star ruby, star sapphire, some garnets, for example – have the extraordinary ability to cause light to appear as star-like rays radiating from the centre of the stone. This property is called asterism.

Although these optical properties are inherent in gemstones, their full effects can only be realized when the stones are shaped and polished. Cutting and shaping gemstones in order to achieve the best angles between the faces of the stone, is an expert job. Polishing the faces is also important, and helps to enhance the light-catching properties of the stone.

The second necessary quality for a mineral to possess before it can be classified as a gemstone is the quality of durability. Gemstones must have this property for two main reasons. First, gemstones are used for decorative jewellery, and therefore undergo a good deal of 'wear and tear'. Minerals that have high hardness values – 7 or more on Mohs' Scale of hardness (see page 30) – are the best equipped to withstand normal usage. Other gemstones – opal, turquoise, lapis lazuli and amber, for example – have lower hardness values on Mohs' scale, and thus merit special care during use. The second reason why durability is such a necessary property is that cutting and polishing are usually required in order to realize the stone's gem-like qualities. Soft stones would be ground down to powder during these processes. Usually only non-cleavable minerals are considered as gemstones, since during the cutting

LEFT **Beryl** The already cut and shaped semi-precious gemstone in its usual yellow colour. Pure beryl is colourless. The yellow colouration is due to the presence of traces of iron impurities. Aqumarine is also a form of beryl, containing traces of iron. The semi-precious pinkish gemstone morganite is beryl which contains traces of lithium.

BELOW **Emerald** Two beautiful examples of the six-sided (hexagonal) emerald crystals, before cutting and polishing. Emerald is the most precious of the beryl family. Their green colour is due to the presence of chromium impurities. Emeralds are always mined from the parent rock, which is exposed at the surface through weathering processes.

RIGHT **Malachite and azurite** A polished section of the copper-bearing minerals, malachite and azurite. In this photograph, the malachite is green, the azurite is blue. When cut and polished in this way, malachite reveals an intricate and beautiful pattern. This makes it one of the most decorative and most sought-after of the semi-precious gemstones.

processes, stones can easily split along readily cleavable planes and thus be ruined. But there are exceptions. Diamond, emerald and topaz are examples of gemstones which have planes of cleavage. Expert cutters can, however, utilize these planes in the cutting processes – one of the skills which make gemstone-cutting such an expert job.

The third property, rarity, is the result of man's acquisitive instinct. Rare and scarce materials are always sought after, and thus acquire a relatively high value. The rarity of a gemstone may be real or promoted. The real scarcity, for example, of Colombian emeralds and Burmese rubies, is an important factor in their status as precious stones. On the other hand, diamonds are not particularly rare, but their value and status is maintained by the strict control of their output. Most diamonds are small in size and inferior in quality, and are used for industrial purposes.

Most gemstones are true crystals, but some may be simply pieces of rock. Lapis lazuli, for example, is a rock composed of the mineral havynite, with golden flecks of pyrites and white wisps of calcite, but it fulfils the criteria of the gemstone family and is classified as a gemstone. Obsidian, the black glassy igneous rock formed by rapid cooling of liquid magma is also considered to be a gemstone.

The gemstone family also includes minerals of animal origin. The pearls found inside the shells of oysters and mussels are not naturally occurring minerals, but are formed by the secretion of natural fluids. Some coral is occasionally used for gem ornamentations and is another example of an organic gem of marine origin. Gems of plant origin include amber, which is the fossil sap of conifer trees, and jet, which is a variety of the brown coal known as lignite.

Modern science has also created its own family of gemstones. Glass can be moulded and coloured by the addition of chemicals into imitation gemstones. The general name given to such imitations is *paste*. It is possible to make some synthetic gems from the same materials that formed the true gems, but this is often a lengthy and expensive process. For example, man can make diamonds! The comparatively poor gemstone quality of the resulting stones makes the process of

industrial interest only. Existing gem minerals can be changed in colour by the effects of heat, radioactivity and the addition of chemicals. For example, the common brown zircons may be changed into blue, yellow or transparent zircon gems by heating. A similar process changes dull smoky quartz into the more desirable gemstone known as yellow citrine quartz. Green diamonds may be produced by subjecting transparent diamond to radioactivity.

Gems collected 'in the field' will be true mineral gemstones, but gems bought as jewellery are more difficult to classify. The amateur geologist will find it almost impossible to distinguish natural gems from artificial or artificially improved gems. However, price will often give an indication of their true nature.

Cutting and polishing techniques

Although there are some inexpensive kits designed for the amateur geologist and suitable for the cutting and shaping of gemstones, these techniques are usually best left to the experts, especially if the gemstone is of some value. Some of the simpler cuts however can be made by the enthusiast on the semi-precious gemstones. Pearls are the only gemstones that can be used in their natural state although often they are 'skinned' by a 'pearl doctor'; all others need cutting and polishing before their true beauty is revealed. There are

RIGHT **Tourmaline** This coloured tourmaline crystal is embedded in quartz. It will be cut and polished into a semi-precious gemstone, and perhaps become earrings, a ring or a pendant.

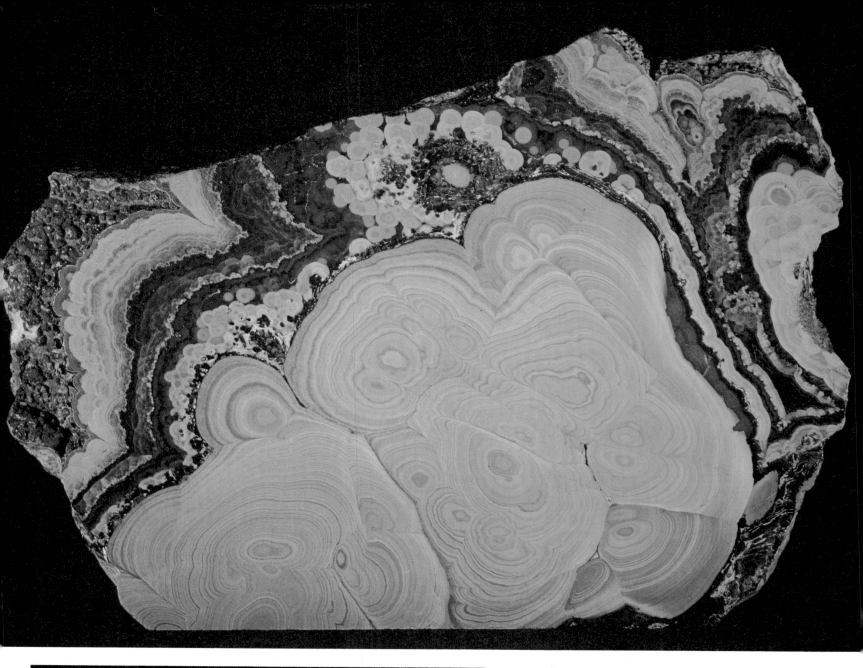

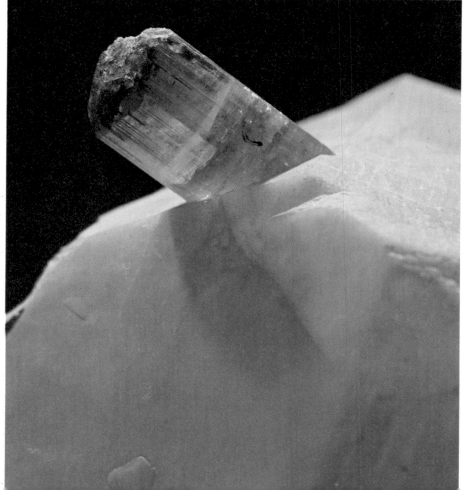

many different cuts, largely determined by the optical properties of the stone in question. The simplest style of cutting, known as *cabochon*, was known to the early Chinese and to the ancient Egyptians. It is still probably the most popular cut for most gemstones. Cabochon consists of cutting a circular, oval or lens-shaped gem with a domed roof of high or low curvature. The gem may be completely rounded but for setting in rings and pendants the bottom of the gem is generally flattened. Gems which depend on their interesting colours and patterns, such as the opals, cat's eyes, agates and turquoises, are best fashioned in the cabochon manner. Starstones of ruby and sapphire are also generally of this cut.

The style of cutting in which the gem surface is flattened is known as *faceting*. The number of faceted surfaces varies – only a few faceted surfaces may be cut, or very intricate faceted patterns may be achieved. The styles of faceting are numerous and depend both on the type of gemstone to be cut and on the condition it is in. The art of faceting can be utilized to remove unwanted flaws or irregularities in the gemstone. The outline shape of the gemstone may be round, oval, rectangular or square. The rose-cut of 12 or 24 triangular shaped facets terminating in a point and the brilliant-cut (the modern cut for diamonds using 58 facets) are two types of cut which can be achieved

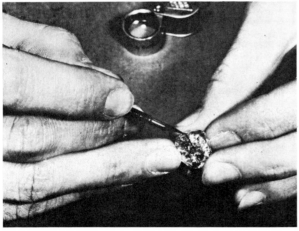

master. Mounting your tumble-polished gem material is a natural culmination of the polishing processes. The amateur can easily polish and mount semi-precious gemstones and minerals to make a wide variety of jewellery for display or for gifts.

The occurrence of gemstones

Gemstones occur in all three of the major rock groups, igneous, sedimentary and metamorphic, and the same gem species may occur in all three. Igneous rocks yield the greatest variety of gem species; the granite family is the most prolific in variety of gems. This richness is a result of the concentration of volatile gases and liquids present in the last stages of solidification of the magma, when water and rare elements

FAR LEFT **Examination** The expert examines the diamond to decide how the stone shall be cut. He marks the diamond with Indian ink along the direction in which he wishes to cleave it.

with rounded gems. Oval-shaped gems are generally faceted in the *briolette* style with the entire surface cut in triangular or rectangular facets. Rectangular and square gems are usually faceted with the *scissor* or *step cut*.

Polishing is an inexpensive and simple method of improving the quality of some of the semi-precious gemstones as well as other hard minerals not generally considered to be of gem quality. Attractive results can be produced by tumbling small pieces of mineral (or even rock specimens) in a specially designed rotating barrel, which contains first abrasive powder, and finally polishing powder. Polishing is a rewarding hobby for the mineral collector and since instructions come with the polishing kits, the process is easy to

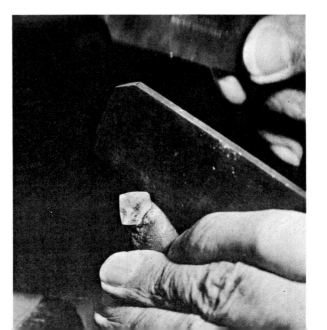

LEFT **Cleaving** The rough diamond is transformed into a brilliant gem by a number of precise operations, each designed to allow the diamond to give the maximum refraction and dispersion of light. After the stone has been carefully examined it is ready for cleaving. Here the operator is about to divide the stone along its grain (cleavage). Any mistake at this point may result in the shattering of the stone.

FAR LEFT **Sawing diamonds** A diamond is sawed when it must be divided across its grain. It is set in solder into a metal dop which is then clamped into an arm above the saw so that the blade will cut along the line marked on the diamond in Indian ink. The thin phosphor-bronze blade has its edge impregnated with diamond dust. As it revolves against the diamond, which is held onto it by gravity, it continually impregnates itself from the stone it is sawing. Although the saw runs at high speed, it can take hours to cut through a small stone.

LEFT **Bruting** A diamond is used to abrade the one which is to be polished.

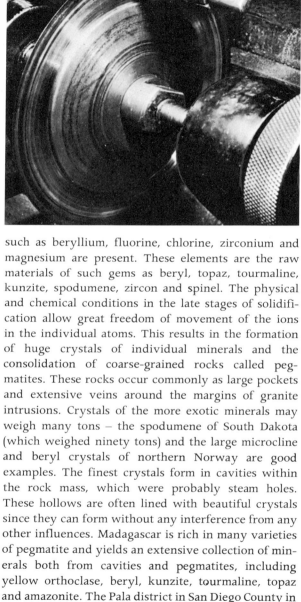

such as beryllium, fluorine, chlorine, zirconium and magnesium are present. These elements are the raw materials of such gems as beryl, topaz, tourmaline, kunzite, spodumene, zircon and spinel. The physical and chemical conditions in the late stages of solidification allow great freedom of movement of the ions in the individual atoms. This results in the formation of huge crystals of individual minerals and the consolidation of coarse-grained rocks called pegmatites. These rocks occur commonly as large pockets and extensive veins around the margins of granite intrusions. Crystals of the more exotic minerals may weigh many tons – the spodumene of South Dakota (which weighed ninety tons) and the large microcline and beryl crystals of northern Norway are good examples. The finest crystals form in cavities within the rock mass, which were probably steam holes. These hollows are often lined with beautiful crystals since they can form without any interference from any other influences. Madagascar is rich in many varieties of pegmatite and yields an extensive collection of minerals both from cavities and pegmatites, including yellow orthoclase, beryl, kunzite, tourmaline, topaz and amazonite. The Pala district in San Diego County in Southern California has rich violet lithia mica pegmatites with multi-coloured tourmalines, pink beryl, and kunzite. Beautiful crystals of tourmaline were obtained from Mount Mica in Maine in the latter half of the nineteenth century. Two young men, Elijah Hamlin and Ezekiel Holmes, found the tourmaline crystals weathered from the parent rock and lying

loose on the surface of the ground. This find led to extensive exploitation of the tourmaline-rich cavities present in the igneous rocks of this area. Large quartz veins often branch off from the margin of the granite intrusions. These are associated with the pegmatite stage of the cooling magma and often contain gold. They also hold fine crystals of garnet, amethyst, tourmaline and topaz. The intermediate igneous rocks, particularly the corundum syenites, contain rich ruby corundum. Unfortunately only a small proportion of it is of gem quality. The ultra basic igneous rocks, such as peridotite, are rich in olivine (peridot) but again gem-quality peridot is rather rare. The kimberlite (mica peridotite) pipes of South Africa are the source of much of the world's diamond production.

ABOVE **Diamond rings** Diamonds may be cut in a variety of styles, the three most popular being the marquise, the emerald cut and the brilliant. These cuts are shown respectively in these three rings.

BELOW **Aquamarine** This crystal owes its colouration to the presence of iron impurities, and is a form of the mineral beryl. An aquamarine crystal found in a pegmatite vein in Brazil in 1910 measured nineteen inches by sixteen inches. Pliny once said that aquamarines were stones 'which imitate the greenness of the clear sea'. Brazil and Sri Lanka are its chief sources.

BELOW RIGHT **Blue topaz** This crystal shows the typical shape of the blue topaz crystal. It is still embedded in the matrix. The blue variety is the most prized of all the numerous shades in which topaz is found. Blue topaz is sometimes mistaken for aquamarine, but it has a greater specific gravity.

The sedimentary rocks provide some excellent gemstone material which has been deposited by percolating waters through joints, fissures and intergranular spaces in the rocks. The high-quality emeralds of Colombia in South America probably originated as mineralized waters passed through the black shales and sandstones of the Cretaceous period. It is interesting that Colombian emeralds occur in two main localities, Muzo and Chevos, and that they are distinguishable from each other by the presence of other minerals in the gems. The former contain dark organic material, and the latter small crystals of iron pyrites. This kind of 'identity mark' is often present in minerals. It may be used to differentiate between stones from different localities, or may characterize a specific single cut gemstone. Another point of interest arising from the Colombian emeralds is that good gem material is difficult to obtain from areas of tectonic disturbance – in both the Colombian areas intense folding, faulting and thrusting have disturbed the emeralds along the cleavage planes and shattered the gemstone material. Opal is another gemstone deposited by percolating waters and fills the cracks and intergranular spaces in the sandstones and limestones of the Australian occurrences. Opal is also found in the volcanic rocks of Slovakia. Turquoise was laid down by percolating waters in the lower Carboniferous rocks of the Sinai Peninsula.

Metamorphic rocks yield ruby, beryl, garnet, lapis lazuli, jade and emerald, among many other gemstones. The changing physical and chemical conditions, resulting from metamorphism, induce the rearrangement of the ions in the atoms and gem minerals such as garnet, ruby and emerald are formed. In the Mogok Stone tract of Upper Burma the metamorphozed limestones contain the rich pigeon-blood coloured rubies, sapphire and spinel. Emerald is found in the mica schists of the Austrian Tyrol. The lovely green peridot occurs in association with the nickel-bearing veins in the dunite (a metamorphozed peridotite) of St John's Island in the Red Sea.

Some species of gemstones can be obtained from more than one type of rock. For example, emeralds can be obtained from both sedimentary and metamorphic rocks; garnets of different types from igneous and metamorphic rocks; rubies from metamorphic and igneous rocks.

Alluvial deposits, which are the accumulations of the unconsolidated products of weathering, are the source of some of the finest gem material. Sometimes, as in the spinels of Burma, the minerals have not travelled far from their source and still retain some of the sharp crystalline outline. Often the crystals have been carried some distance, generally by running water. During this process they have suffered abrasion and lost their angularity, and occur as rounded pebbles. Minerals with pronounced cleavage, such as feldspar, beryl or emerald, tend to be much broken; minerals of low hardness or advanced alteration also tend to be broken down. Diamonds, topaz and garnet become rounded pebbles, often with frosted surfaces. The main consequence of the travel process is to destroy all weak and incoherent material and leave

LEFT **Ruby** The word ruby comes from the Latin *rubeus*, red. Rubies are corundum, a major aluminium mineral, and obtain their red colouring from the presence of minute amounts of chromic oxide. (Stones which depend on impurities for their colour are called *allochromatic*, stones which have colour when pure are called (*idiochromatic*.) The hexagonal form of the crystal, shown here, is typical of the ruby and of its 'sister' stone, sapphire. The ruby is found in many different shades of red, but the most valued is the pigeon's blood red stone (a red that is almost purple). Chemically, the ruby is formed of alumina and oxygen. It has a hardness of 9 on Mohs' scale of hardness, diamond being the only gemstone with a greater hardness. The ancients believed that the ruby carried the spark of life. Henry V carried the Black Prince's ruby at Agincourt, for fortune. Perhaps it is as well he did not know that it was not a ruby after all, but spinel, often mistaken for ruby.

only the best gemstones in the gravels. The gem gravels of Sri Lanka, formerly Ceylon, are among the richest in the world. They are derived from a complex of igneous rock together with schists, gneisses and metamorphozed limestones. The list of gems from the Sri Lanka gravels is very long but alexandrite, sapphire, ruby, beryl, garnet and spinel are among the finest obtained from this source. Madagascar and Vietnam are also fruitful sources of alluvial gem material.

The weight measure of all precious stones is the *carat*, one carat being one fifth of a gram.

Diamond

Of all the gemstones, diamond is undoubtedly the most desired. It has a history steeped in myth, legend and incredible tales of violence, greed and intrigue.

LEFT **Diamond** A rough diamond before cutting, set in the 'blue-ground' which has been blasted from the volcanic neck far below ground.

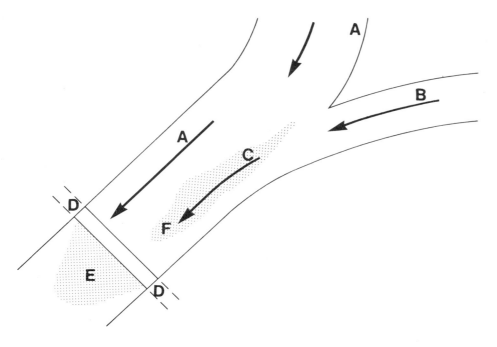

Yet the diamond is not the most valuable of the gemstones, for certain rubies and sapphires are more rare and costly.

Diamond is composed of pure carbon atoms – the same material as graphite in a pencil, and coal in a hearth. It is the way in which the carbon atoms are arranged that makes diamond so distinct from other members of the carbon family. The arrangement of the atoms is responsible for the unique quality of hardness which diamond possesses, for diamond is the hardest of all naturally occurring substances. It is also responsible for the reflecting and refracting properties which give the diamond 'the radiance of the sun, and the splendour of the rainbow'.

Most modern diamonds come from the South African diamond mines, or from placer deposits weathered from them. Two of the most famous diamonds, the Koh-i-Nor and the Blue Hope, were probably found in or near the once-famous diamond mine of Golconda in India. These mines have long been worked out however, and until diamonds were discovered in South Africa in 1850, Brazil was the world's major diamond producer. Her most important mines are near the town of Tejuco, now called Diamontine. Diamonds are also found in many central African states and, on a limited scale, in Arkansas in the United States, the Croghan Mountains in Ireland and, according to legend, on Ben Hope in Scotland.

Diamonds are generally colourless gems, but they can be blue, yellow, green, pink, red, brown, violet and even black. The deep yellow Tiffany diamond is probably the most famous coloured diamond, although there are other well-known examples. A 24-carat pink diamond was presented to the present Queen Elizabeth II in 1947 by Dr Williamson, a Canadian geologist, and is estimated to be worth £450,000. The red Paul diamond is part of the Russian crown jewels, and the Dresden diamond, once in the possession of Augustus the Strong, is apple green in colour.

ABOVE **Placer deposits** One situation in which placer deposits may occur is in the confluence area of two streams and in the vicinity of a lode crossing a stream. The direction of flow of the main stream is marked by the arrow A, and the direction of flow of the tributary by arrow B. At C the flow is obstructed, and the placer deposit is formed at the stream junction F. DD marks a gold lode and E is the downstream deposit from the lode.

ABOVE RIGHT **The Koh-i-nor Diamond** This famous diamond, mined in Golconda in India was originally 186 carats. Queen Victoria had it recut down to 108.93 carats in 1862 to make it more brilliant.

BELOW RIGHT **La Belle Hélène Diamond**, mined in South Africa. Once 160 carats, it was cut to yield three flawless gems: two matching pear shapes of 38.38 carats and 29.71 carats and a marquise of 10.50 carats.

Most of the diamonds found are not of gem quality, but because of their superlative hardness they have a range of common industrial uses, particularly in cutting, grinding and drilling machinery. The diamond is the national gemstone of South Africa, Britain and Holland – the latter two countries containing the two most important diamond-cutting and setting centres, London and Amsterdam.

There are many famous individual diamonds, each with its own peculiar and fascinating history.

The Koh-i-Nor Few gemstones have had such a long and romantic history as this diamond. It was probably found some 5000 years ago by a shepherd in Maharashtra in India, and it has been calculated that the original uncut stone weighed about 800 carats. The authentic history of the Koh-i-Nor dates from 1304, at which time it belonged to the Rajah of Malwar. A neighbouring prince stole the diamond in conquest but in 1526 his descendants lost it to Sultan Baber, who led the Moguls in their wars against the Indian states. The diamond is said to have formed one of the eyes of the famous gem-encrusted 'Peacock Throne' in the audience hall at Delhi. This fabulous throne is described by the French jeweller, Tavernier, who catalogued much of the gem wealth of the Indian princes. The Koh-i-Nor remained in Delhi until the Persian conquest in 1739 when, after sacking the city, Nadir Shah took the Peacock Throne, together with the diamond, back to Persia. According to legend, Nadir Shah obtained the jewel by trickery. He learned that the Mogul Sultan had hidden the diamond in his turban. He invited the Mogul to a great feast and there asked him to change turbans. The turban was unwound and the jewel was revealed. The Nadir is said to have exclaimed 'Koh-i-

ABOVE **The Eureka Diamond** after cutting and polishing. This was the first diamond ever found in South Africa, hence its name. It is 10.73 carats.

LEFT **The Niarchos Diamond** Cut and mounted it is 128 carats, though it was 426.5 carats in the rough. It also gave two other stones, an emerald cut of 40 carats and a marquise of 30 carats. Mr Stavros Niarchos owns the largest.

nor', which means 'Mountain of Light'. In Persia the diamond seems to have brought ill-luck to its successive owners, and this period of its history contains many stories of death and destruction. The stone was eventually recovered by Ranjit Singh, the 'Lion of the Punjab', and on his death was placed in the Punjab treasury in Lahore. In 1849 the East India Company annexed the diamond in payment for the damage and destruction caused by the Sikh revolt. It was then presented to Queen Victoria and recut at a cost of £16,000. During the cutting process it was reduced in size to 109 carats. Subsequently it has graced the crowns of Queen Victoria, Queen Alexandra, Queen Mary and Queen Elizabeth, the Queen Mother. Legend has it that the Koh-i-Nor brings ill-luck to male wearers, but good luck to a female wearer, and history seems to support this view.

The Idol's Eye Some diamonds have remained in private ownership yet have equally fascinating histories – the Idol's Eye is one such gem. It was mined at the Golconda Mines in India in 1600 AD. It was cut into a rough pear-drop shape by an unknown Indian craftsman, who cut enough facets to highlight its beauty. After cutting, its weight was 70.2 carats. The original owner was the Persian Prince Rahab. He managed to get deeply into debt with the East India Company and eventually they took the diamond in payment. At this point the stone seems to have disappeared and for almost 300 years its whereabouts remained a mystery. It reappeared in the Turkish temple at Benghazi in the ownership of Sultan Abdul Hamid II. He became short of money and sent the stone to Paris in order to sell it. Sultan Abdul Hamid was obviously a poor judge of men, for the messenger entrusted with the task staged a robbery and the stone was next seen in a pawnbroker's shop in Paris. Before the Sultan could claim the stone it was bought by a Spanish nobleman. He placed it in a bank in London where it remained until the end of World War II. In 1947 it was bought for 600,000 dollars by Harry Winston, a New York jeweller. It was later bought by the Chicago jeweller Harry Levenson, who insured it for 1,000,000 dollars.

The Cullinan Diamond The Cullinan is the largest diamond ever found. It was found by accident in the Premier Mines near Pretoria in the Transvaal by an employee on an inspection tour, who literally stumbled across it. In its uncut state it weighed 1,106 carats (over $1\frac{1}{4}$ lbs). The stone showed one fine cleavage face and it is almost certain that a large part of this stone had been lost in the mine. The imagination baulks at the thought of the original size of the stone! The stone was named after Thomas Cullinan, chairman of the mining company. In 1907 it was bought by the Transvaal Government, who presented it to Edward VII. So large a stone must be carefully cut and polished and this work was entrusted to a famous diamond cutter in Amsterdam – J.J.Asscher. He decided in the first instance to cleave the diamond into three portions. His responsibility was great, for if he mis-selected the plane of cleavage the crystal could shatter into small fragments when struck. Asscher spent a great deal of time in selecting the potential plane of cleavage. He carefully placed the blade and struck – nothing happened! Asscher struck again and the diamond parted cleanly along the cleavage. The strain had been too great –

Asscher fainted and had to be revived with brandy. Eventually nine large stones and ninety-six smaller ones were cut from the Cullinan diamond. The four largest stones were named: the Star of Africa, 530 carats, reputed to be the largest and most perfect diamond in the world, is now in the royal sceptre of the British Monarch. The Second Star of Africa weighs 317 carats and is in the Imperial State Crown of England. The Third Star of Africa, 94 carats, and the fourth stone, 63 carats, were both put in the Coronation Crown of Queen Mary.

The Hope Diamond The Hope is another diamond of ill-omen. It is deep sapphire blue in colour and weighs 44 carats. The original larger stone was bought in 1642 by the French jeweller Tavernier from the Kollur Mines in India. He later sold it to Louis XIV. It was named the Tavernier Blue and became part of the French crown jewels. During the French Revolution in 1792, the Tavernier Blue was stolen with the other crown jewels and never recovered. In 1830, however, a deep sapphire-blue, brilliant-cut stone appeared in London. It weighed 44 carats and it became apparent that the original Tavernier Blue had reappeared, although it had been cut in the interim period. It was bought by Thomas Hope, an American banker, who later sold it to the Sultan of Turkey for £80,000. In 1911, Harry Winston of New York, who also owned the Idol's Eye, bought the gem and presented it to the Smithsonian Institution in Washington, where it now rests. A curious incident occurred in 1874 when a smaller diamond, identical in colour to the Hope, appeared at the sale of the Duke of Brunswick's gemstones. Allowing for the loss of small amounts of material in cutting, the weight of this stone added to the 44 carats of the Hope Diamond, would be 67 carats, the weight of the original Tavernier Blue. Although there is no real proof that both of the stones had been part of the original blue diamond, this is certainly a most extraordinary coincidence!

BELOW **Imperial State Crown** Part of the British Crown Jewels. This crown together with the other magnificent regalia which forms the British Crown Jewels, is kept on public display in the Tower of London. Every year many thousands of visitors from all over the world quene, sometimes for hours, to see this magnificent collection of precious gemstones.

ABOVE The Shah's Crown
Part of the magnificient
Crown Jewels of Iran, this
crown weighs over 2,000
grams and consists of gold,
diamonds, emeralds, sapphires
and pearls.

RIGHT The Peacock Throne
Part of the Crown Jewels of
Iran, the Peacock Throne
contains 26,733 jewels. The
Crown Jewels of Iran back
the nation's currency.

Pearl earrings
Matched pearl earrings,
mounted with gold and
diamonds.

Ruby and Sapphire

These stones are both varieties of the normally drab
grey mineral, corundum. They have identical properties to corundum, differing only in colour. Ruby is red
corundum (Al_2O_3) and contains chromic oxide impurities. Sapphire is generally cornflower-blue in
colour, although green, purple, yellow and pink
varieties can be found. Ruby and sapphire often occur
together in the same parent rock, since their geological
origins are similar.

**Taking pearls from
the oysters, Cape York,
Australia** The oyster shells
have been brought to the
surface, where they are
opened and examined for
pearls. Only a small percent-
age of shells contain pearls
of value.

The ruby is sometimes known as 'Lord of the Gems',
and was considered by the Romans to be the gemstone
of Mars, the god of war. In colour it resembles a blazing
red sunset, and because of this was thought to signify
command and victory in battle. In the fourth century
AD St Epiphanius of Constantia wrote that rubies
shone in the darkness, and contended that they could
shine through clothing with undiminished fire. Compared with other gems, there is little folk-lore attached
to rubies, a fact probably attributable to the rarity of
perfect ruby gemstones. The largest natural ruby is
thought to be one found in Burma which weighed 1184
carats. In 1961, a broken ruby was found in the United
States. The largest piece weighed about 750 carats, but
the total weight of the pieces was 3421 carats. Labora-
tory-made ruby prisms, used for laser technology, are
sometimes more than twelve inches in length. Two of
the most famous 'rubies' – the Black Prince's Ruby and
the Timur Ruby, both in the British crown jewels – are
not rubies at all! They are red varieties of the gemstone
spinel.

The sapphire, according to tradition, is associated
with intelligence. The Ancient Greeks dedicated the
stone to Apollo, and Alfred the Great thought it cured
abscesses and headaches, 'brought about peaceful
agreements', 'confirmed the mind in goodness' and
'induced piety'! Both Queen Victoria and Alexandra,
the last Czarina of Russia, owned crowns studded with
sapphires, and Queen Elizabeth's II's crown carries the
famous Stuart sapphire. The largest sapphire in
existence is the Black Star sapphire of Queensland,

which weighed over 2000 carats uncut, and from which a bust of President Eisenhower was carved. A smaller bust of Abraham Lincoln was carved from an even larger stone, also found in Queensland. The gem gravels of Sri Lanka were once rich sources of both rubies and sapphires, although the supply of rubies seems to have dried up. The 'Hill of Precious Stones', in Thailand, and Burma are other sources of these precious stones. The supply of the famous Burmese rubies also seems to have dried up – one of the main reasons for their rapid increase in value over the last few years. Weight for weight, some rubies are now more valuable than diamonds, emeralds or sapphires, attaining a price of up to £4000 ($10,000) per carat.

Emerald and Aquamarine

Emerald is a green variety of the mineral beryl, the colouration resulting from the minute quantities of chromium which are present. The gemstone aquamarine, which is sea-green or blue in colour, is a close cousin of the sapphire. Its colouring is the result of iron impurities. These two stones have different geological origins and are therefore not found in the same localities. Brazil and Sri Lanka are the chief source areas for aquamarine, while emerald is obtained from the Ural Mountains in Australia, Norway and South Africa, and most significantly, from Colombia in South America. Colombian emeralds are associated with the history of the Incas and formed part of the loot of the Spanish Conquistadors. The last Inca king possessed a head-dress studded with 500 emeralds, but this was finally broken up in the United States in the 1940s. The largest known emerald in the Devonshire Emerald, which weighs 1350 carats and is nearly a foot in length. This is a Colombian emerald and is at present in the British Museum in London. Gemstones related to emerald are heliodor, a yellow variety of beryl; morganite and goshenite, which are pink and colourless varieties respectively.

Pearls and semi-precious stones

Pearls are gemstones which form perfectly in the natural state and require no cutting or polishing to

BELOW LEFT **Cutting opal, Cobber Pedy, Central Australia** 95% of the world's opal deposits are found in Australia. The opals are cut and polished by hand, and then they are mounted to make rings and other jewellery.

ABOVE **Mining for opals at Lightning Ridge, New South Wales** Opals often occur as small nodules in a soft clay layer beneath sandstone. The picture shows an opal miner working 90ft below ground.

BELOW **Carvings in jet** Jet is a hard black substance of organic origin and is not strictly a gemstone. However it can be carved into intricate patterns and representations for jewellery. It was very popular in Victorian times.

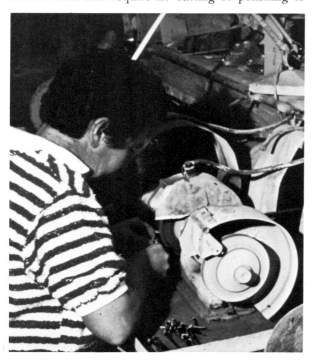

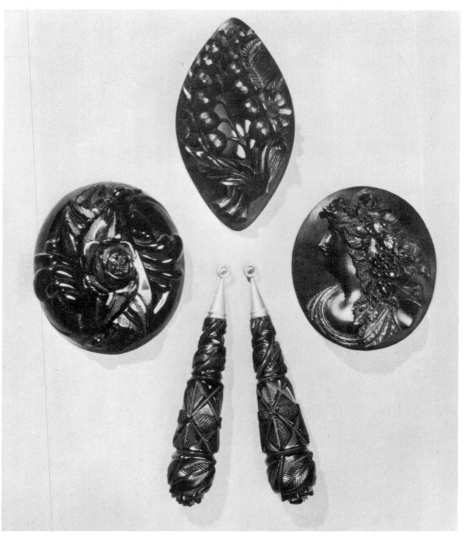

reveal their true beauty. Pearl is not a mineral in the strict sense of the word, since it results from a solution secreted by certain of the bivalved molluscs – the pearly oysters and the pearly mussels – and is therefore organic in origin. Pearls were probably the earliest gemstones, and nowadays rival diamonds as the most sought after. Pearl-fishing has taken place for more than 2000 years in the Persian Gulf and in the Mannar Gulf off the north-west coast of Sri Lanka. More recent source areas include the inshore waters of north and west Australia, the Pacific Islands, the Caribbean and the Gulf of Mexico. Pearl-bearing molluscs are also found in several rivers in the Mississippi drainage system in the United States. The British crown jewels include pieces set with pearls that were taken from British rivers and coastal waters.

Pearl, like all other organic matter, decays sooner or later, and so there are few examples of famous, historic pearls. An exception is that known as La Peregrina – 'The Wanderer' – which has been worn by many famous and wealthy people, including King Philip II of Spain, Mary Tudor and, more recently, the actress Elizabeth Taylor. One of the largest recorded pearls, reputed to be the size of a pear, was given by Shah Jahan to his wife. He also built the Taj Mahal for her. The largest known natural pearl is the Pearl of Lao-tze,

LEFT **The Murchison Snuff Box** This gold box is inset with 16 large diamonds and many smaller ones. The enamelled picture in the centre is of Tsar Alexander II who presented the box to Sir Roderick Murchison in 1867. Murchison was an eminent geologist who worked extensively in Russia.

also called the Pearl of Allah, which measures $9\frac{1}{2}$ inches in length and $5\frac{1}{2}$ inches in diameter, and weighs more than 14 lb. It was discovered in the shell of a giant clam at Palawan in the Philippines in 1934. In 1939 this pearl was valued at £1,250,000 ($3,175,000).

Pearls are formed when an irritant – a grain of sand, for example – enters the shell and is covered by layers of the mineral calcite which is secreted from special glands within the mollusc. This natural process can be stimulated artificially, by implanting irritants inside the shell of the mollusc. The Japanese are particularly good at this, and produce thousands of *cultured* pearls each year.

Most natural pearls have a misty grey lustre, but they can be coloured and pink, green, blue, yellow and black pearls have been found.

Garnets contain a wide variety of impurities, such impurities giving an array of colours to these popular and inexpensive gemstones. Garnet is a common metamorphic mineral, but most examples are small and worthless – although they provide very attractive gem material for the amateur geologist. They are mostly used in industry as abrasives (having a hardness value of 7 on Mohs' scale), and as the 'jewels' in inexpensive watches. The most important source area for gem garnets is Bohemia in Czechoslovakia, where most of the garnets common in Victorian jewellery were mined. Garnet is, in fact, the national gemstone of Czechoslovakia. Garnets are generally considered to be lucky stones, and traditionally ensure a successful life and freedom from accident for the wearer.

Tourmaline is another commonly found gem mineral and is associated with the igneous granite regions of the world. Like garnet, tourmaline is found in a wide variety of colours, a single crystal often showing two or more colours. Brazil, Sri Lanka, Madagascar and South Africa are the most important source areas. Many people prefer tourmaline to the gemstone opal, which it resembles, both for the attractiveness of its colouring and for its relative cheapness. Tourmaline traditionally brings peace and serenity to its wearer, which is perhaps another reason for its popularity.

Gemstones belonging to the quartz family include the precious gemstone, opal, and numerous semi-precious stones. Agate, amethyst, bloodstone, cornelian, citrine, cat's eye, tiger eye, rose quartz, onyx and jaspar are all members of the family. These semi-

ABOVE **Lapis Lazuli bowl**
Lapis lazuli is not really a gemstone, but a rock. Because of its beautiful blue colour (the mineral azurite) and the gold pyrite flecks and white calcite, it is often classified as a gemstone. Lapis lazuli is fairly soft and can be formed into vases, bowls and other ornamental pieces. Catherine the Great of Russia had the walls of a ballroom entirely made of lapis lazuli.

LEFT **Butterfly brooch set with polished opals** Many opals, including the highly prized black opal, are dug from the soft clay layers of Lightning Ridge, Australia. The cutting and polishing of opals is a very delicate and highly skilled operation. The opals are cut with a diamond-studded saw polished to bring out the natural colour and can be used to make a wide variety of jewellery.

precious stones have the same properties as colourless or white quartz, and can all be used to make attractive and decorative jewellery.

Opal The most precious gemstone in the quartz family is opal. It is slightly softer than quartz and contains vivid patches of colour set against a milky white background. It is this quality that makes opal such an attractive and sought after gem. The richness in colouring is a result of the geological origins. Small amounts of silica jelly, derived from hot spring waters containing different quantities and types of mineral, are present in the stone. Since the beginning of this century, New South Wales, Queensland and South Australia have been the most important source areas for opals, although they also occur in Mexico, Hungary, Nevada in the United States, and many other localities.

Despite its attractive appearance and popularity, opal is considered to be an unlucky stone, since it can easily lose its sparkle and sometimes splinters. Many opal-wearers in the past – the short-lived Prince Rudolph of Hungary and King Alfonso XII of Spain, for example – have had unhappy lives, which has also helped to promote the tradition of ill-luck.

Topaz and Zircon are gemstones which in their transparent state are often mistaken for diamonds because of their high hardness values and their 'fire'. Both stones occur in a variety of colours. Blue topaz is often mistaken for aquamarine, yellow topaz for citrine quartz – which demonstrates that colour is an unreliable characteristic on which to base an identification of mineral or gem material. Topaz is a rare stone and therefore a gemstone in its own right, the colourless and blue stones being the most sought after. Colourless varieties are found in placer deposits in Brazil and in the eastern areas of the United States. Japan and Sri Lanka provide most of our blue topaz gems.

The blue and colourless varieties of zircon are also the most popular. The most important source areas the placer deposits in Ontario and Quebec in Canada, and in New York and New Jersey in the United States. Zircons are also found in Fort William and the islands of Harris and Tiree in Scotland. Commercial production is mostly in the south east Asian countries. Zircon, like calcite, has the property of double refraction, a property which allows for easy identification, regardless of colour.

Lapis Lazuli is always found associated with limestone which has been intruded by molten igneous rock. It is opaque, which makes faceting unnecessary. This beautiful deep blue mineral is yet another gem material with a long and fascinating history. Specimens have been found in Hittite tombs dating from the seventh century BC, and in the tombs of the Ancient Egyptians. The Sumerians believed that anyone who carried lapis lazuli as an amulet carried a god with him! The most important source area is at Badakshan in Afghanistan, where the closely-guarded mines have been worked for thousands of years.

The gemstones mentioned in this chapter are among the most beautiful and sought after, but there are many other gemstones to add to a collection – spinel, moonstone, alexandrite, amber, peridot, jade, jet and turquoise are a few. The story of gemstones forms one of the most fascinating subjects in geology.

Map of World Gemstone Deposits

◇ Diamond	● Ruby
✳ Sapphire	◠ Amber
▣ Emerald	▼ Garnet
○ Opal	▽ Peridot
☐ Quartz	▦ Pegmatite
▲ Amethyst	◔ Cordierite
✸ Chalcedony	▧ Jade
◠ Malachite	☐ Lapis Lazuli
◉ Rhodochrosite	
☆ Turquoise	
● Zircon	

Chapter 7
MINERALS, ROCKS AND MAN

From the earliest days of history rocks and minerals have played an important role in man's survival and development. Our remote ancestors relied on natural rock caves to give them shelter and protection. Today man uses natural rock to build his houses and factories. Stones and arrow heads made from flint and obsidian were the first lethal weapons. Today we have more 'sophisticated' weapons which utilize minerals such as uraninite, carnotite and autunite. For thousands of years man has used mineral fuels to keep him warm, to bake his pottery and to smelt his ore minerals. He has used gemstones to ornament his body and his clothes. The minerals and rocks can truly be named the 'fruits of the earth', and the names we have given to ages in man's history indicate their importance in his development – the Stone Age, the Bronze Age and the Iron Age.

Generally speaking we can sub-classify minerals into two groups – the metallic minerals and the non-metallic minerals. The metallic minerals are those from which we can derive metals – they are sometimes known as ore minerals. The non-metallic minerals include the sands, clays and rocks used in chemical and construction industries.

The search for minerals

In the past, prospecting for minerals was largely a matter of trial and error. Today we have a number of sophisticated techniques to aid our never-ending search for ore minerals, fuel deposits and precious gemstones. Careful study of rock formations often yields valuable clues to the presence of ore minerals. Radioactive minerals can be discovered with the aid of the geiger-counter. Buried deposits can often be found by enlisting the aid of the geophysicist and his sophisticated electronic equipment. The latest and most exciting technique is the use of photographs taken from satellites as they orbit hundreds of miles above the earth's surface. These photographs often reveal huge and hitherto unsuspected mineral deposits. Once a deposit is suspected or discovered trial borings are made so that samples can be analyzed, and the size and economic potential of the deposit determined. A good example of this procedure is the trial borings made in the North Sea over the last few years. These borings have revealed the presence of huge oil and natural gas deposits under the sea-bed.

Obtaining the minerals

Once minerals have been located, they must be economically extracted from the earth. There are several ways in which minerals can be obtained once

Mount Isa, Queensland, Australia A magnificent view of the large open cast mining operation at Mount Isa, Queensland. The main mineral mined here is copper. The excavator in the middle of the photograph scoops copper ore into the truck, which will take it to the nearby smelting plant.

ABOVE **The formation of coal** Over many millions of years, layers of carbonaceous debris are compressed.

BELOW **Botryoidal Haematite** This is also known as kidney iron ore from its shape.

the economic viability has been carefully examined.

If the mineral vein, or the ore-bearing rock is close to the surface, mechanical excavators can simply tear off the covering layers of material, revealing the mineral beneath. This process is known as open-cast mining. In the past, worked-out deposits were usually left and formed ugly blots on the landscape, but today many authorities insist that the waste material – the overburden – is replaced into the excavated areas, and the landscape conserved.

Often, rocks have to be broken up or dislodged by blasting with explosives. Much of our slate, marble and granite is obtained in this way. Slab rock of granite and marble is also produced by sawing and wedging. This avoids the development of minute cracks and enhances the value of the rock for use, for example, in sculpture. Such workings are generally known as quarries. Similar workings yield relatively soft materials such as clay, sand and gravel. These are usually known as pits, and the excavation technique is simply to shovel the materials up in huge mechanical shovels.

When a mineral vein or ore-bearing rock is buried at great depths, then it must be mined. Coal-mining is the

most obvious example of this type of excavation. Shafts are dug downwards into the earth until the mineral is reached. Horizontal shafts are then struck out, following the mineral vein or the rock strata. Zinc, copper, salt and diamond are examples of minerals which can be obtained in this way. Some minerals are found already excavated by the natural processes of weathering and erosion. Fragmented material, often rich in metal ores or gemstones, is transported by water and deposited in beds at some point removed from the place of origin. These beds are called placer deposits, and are usually found in

meander sandbanks of rivers, in estuaries and in deltas. The ores or gems are recovered by *panning*, which is a washing process. Dredging is used in many river areas; by this means, large quantities of loose superficial material are obtained for subsequent treatment to obtain the mineral.

Another method of obtaining minerals from the earth is by drilling or pumping. Oil and natural gas are good examples of minerals obtained by drilling. Salt is an example of a mineral obtained by pumping. It is pumped from below the earth's surface as brine, a solution of salt in hot water.

How did the minerals get there?

This is a complex question, but in the simplest terms there are four main processes which result in the concentration of mineral matter. Minerals can crystallize directly from molten magma; they can become concentrated in heated liquids; they can precipitate out of percolating waters, or accumulate from evaporating bodies of water; and they can result from weathering and erosion processes.

Crystallization Minerals which crystallize early in the magma cooling processes are generally of high specific gravity, and may sink to the floor of the magma chamber. They may also be segregated by the action of convection currents within the magma itself. Yet another process involves pressure, where a mush of crystals and magma are separated by a filter press action which removes the magma and leaves the crystals to form an ore body. Concentrations of crystals formed in these ways become bodies of ore, and are usually associated with basic and ultra basic

igneous rocks. Examples of such deposits are the platinum in the norites and peridotites of the Ural Mountains; chromite in the peridotite in South Africa; and copper in the lopolithic gabbro mass and the norites of the Bushveld Complex, also in South Africa. Concentration processes can be sub-classified into three groups – impregnation, metasomatic replacement and cavity filling. *Impregnation* occurs when a concentration of boron, fluorine, phosphorus and minor amounts of other elements are found in superheated waters from cooling magma. This provides a medium in which reaction can occur between the concentrated elements and previously formed minerals. The tin deposits in Cornwall, England and the molybdenum deposits in Australia and Canada are examples of mineral deposits which accumulated in this way. Impregnation is usually associated with acid intrusions, particularly in granite.

As the distance from the intrusion increases so the temperature falls. The result of this is that the deposition of minerals occurs in the order of falling melting points away from the intrusion. This zonal distribution of minerals is illustrated by deposits around granite intrusions in south western England.

It has been shown that the following sequence of ores is met as the distance from the intrusion increases and the temperature falls: cassiterite, wolfram, tourmaline, copper, lead, zinc, iron. The falling temperature also affects the minerals which enclose the ore minerals. Quartz, feldspar and tourmaline characterize zones nearest the intrusions. Fluorite, chlorite and haematite occur further away; while barytes, chalcedony, dolomite and calcite are still further away. Divisions between the zones are not clear cut, so there may be overlapping of mineral types and associations of ore minerals.

ABOVE **Asbestos mine, Canada** A thick sequence of asbestos ore. The ore is dug out of the hillside by huge excavators, loaded onto waiting lorries, and transported directly to the refining plant. After refining the asbestos passes to the weaving mills and is formed into sheets.

TOP CENTRE **Asbestos** A close-up of the mineral asbestos, showing its fibrous nature. Asbestos is a heat-resistant mineral. It can be woven into fabric and made into protective clothing for use by firemen.

Metasomatic replacement is a similar process in which rock is progressively replaced by ore minerals carried in gaseous or liquid media. As the gas or liquid advances it comes into contact with the minerals of the invaded rock which, in the new environment, become unstable. Reaction between the invading media and the rock takes place and ore bodies are formed. For replacement to take place in this way there must be openings along which the mineralizing fluids can travel, and there must be a continuous supply of the mineralizing elements. Examples of ore bodies formed by replacement are found in the iron deposits in the Carboniferous limestone at Llanharry in South Wales, in similar rocks in Cumberland in the Lake District, and in the copper pyrites in Rio Tinto, Spain.

Cavity filling deposits are deposited from solution over a wide range of temperatures and from waters of different origins. Primary openings are those which originated during the formation of the original rock, and include the pore spaces in sandstone and steam holes in volcanic lavas. Examples of ore deposits in pore spaces are often found in the red sandy rocks of the Permian period, as in the basin of the River Don, in Russia and in Turkestan and Germany.

Secondary openings are more important than primary openings for it is in secondary openings that the more valuable ore deposits are usually found. They form subsequent to the formation of the rock in question and are generally sub-classified into two groups – tectonic openings and solution openings. Ore bodies found within the group include fissure veins, saddle reefs, breccia infillings and solution hollows deposit.

In all instances the deposits have a layered appearance. This is a result of concentric deposition – which is rather like the way in which stalactites develop. There is usually an association of non-metallic minerals, such as quartz, calcite or barytes, called the gangue minerals.

Fissures can be formed in many different ways, but usually tensional and shear stresses have played a part. They are often associated with tectonic movements so that they appear in parallel groups or 'en echelon'. The physical character of the rock will determine the reaction it offers to the deforming forces. Almost parallel-sided veins occur, thin and covering large areas, in the hard rocks such as granite or arenaceous sediments. When schists or soft shales are fissured, particularly by shear stresses, the fissures tend to 'pinch and swell'. This characteristic can also be induced by the differential response of rocks to the pressures created by mineralizing solutions. Large volumes of gold, copper, lead, zinc and tin have been obtained from fissure veins. Antimony, radium and cobalt are obtained almost exclusively from such sources. The gold lodes of Cripple Creek in Colorado, the copper deposits of Butte, Montana, and the silver-lead of Prizibram in Austria are a few examples of ores found in fissure veins.

Saddle reefs are formed when there is sharp folding of the rocks, opening a split in the crest of the fold. The quartz-gold reefs of Bendigo in Australia are a good

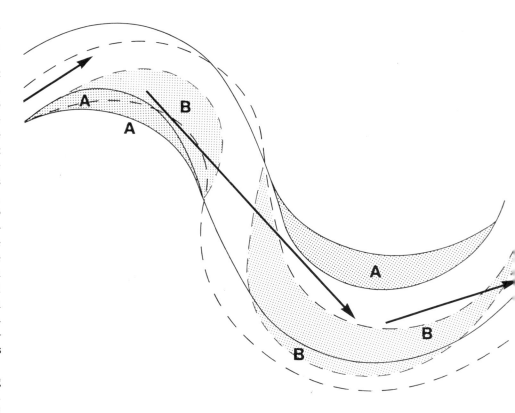

LEFT **Reef gold, New South Wales** The biggest mass of reef gold ever taken from the earth was discovered in 1872 at Hill End, New South Wales, Australia. The mass weighed 630 lb. Its co-discoverer Otto Holter-mann (shown here), used some of the money which he had obtained from the gold to commission Beaufoy Merlin, a remark-able photographer, to take pictures of the gold-fields.

ABOVE **Gold** Gold has always exercised a great fascination over men's minds, yet it is not the earth's most valuable mineral for man's use. There are many historical instances of gold 'fever' – the Klondike and Californ-ian gold rushes, for example–when thousands of people travelled into unknown territories in a usually vain search for this mineral. Today's output is strictly controlled.

LEFT **Meanders and placer deposits** The direction of flow of the stream is shown by the arrows. The limits of the stream as it changes course are shown by the solid and broken lines. A shows deposits along the convex side of the stream in the original position. B shows deposits along the convex side of the stream in the second position. The stream migrates both down valley and across the valley so that the position of the deposits is always changing.

BELOW **Panning for gold, California** A gold prospector panning for gold during the Californian gold rush. Panning was one of the traditional methods used by prospectors for extracting gold. The gold-bearing materials is swirled with water. The heavy gold stays in the pan, while the waste is washed away.

example of this type of secondary opening.

Breccia infillings are commonly associated with faulted zones and are often extremely rich in ores. Many breccias result from a collapse of the subsurface caused by mineral reactions and involving the formation of denser minerals. The copper deposits of Bisbee, Arizona, and the silver, lead and zinc ores of Little Cottonwood, Utah, were deposited in this way.

Solution hollow deposits commonly occur in limestone rocks. Lead, zinc, iron and copper are found in such hollows.

Precipitation takes place when deposited mineral ores are dissolved, carried away in solution and redeposited. The redeposition takes place deeper below the earth's surface than the original area of accumulation, and results in the enrichment of ore bodies at depth. For this reason the process is often known as secondary enrichment. The zone of weathering – the area where the original mineral ores were deposited – is above the water table. Percolating waters 'weather' the ores, chiefly by dissolving them or by oxidizing them. The dissolved minerals are carried downwards by the percolating waters, and the upper areas become impoverished of mineral materials. Because of this impoverishment, the zone is known as the zone of *leaching*. Below this zone and above the water table is

the zone of *oxidized enrichment*. As its name implies, this zone is enriched with oxidized minerals from the higher zones. Immediately below the water table comes the zone of *secondary sulphide enrichment*, where mineral sulphides are found, and below that the zone of *primary sulphides*. Above the areas of secondary enrichment, on the earth's surface, there is often a cap, of varying thickness, called the *gossan*. This is an open, cellular iron-stained mass which sometimes forms a useful indication of the enriched zone below.

The phenomenon of secondary enrichment is illustrated particularly well by an area in Cornwall, England. Acidulated waters percolated downwards and leached the upper portion of the earth, leaving a gossan with stringers of iron and manganese. Chalcopyrite, present in the upper zone of weathering, reacted with the waters to form the soluble sulphate which was carried downwards. Redeposition occured above the water table, and copper oxides and carbonates were formed. Below the water table reaction with the primary copper sulphides resulted in the deposition of bornite and chalcocite. The extent of the enrichment is shown by the figures given for the increase in copper content – the original chalcopyrite contains 34.5% copper, the bornite 63.3%, and the chalcocite 79.8%. Weathering and erosion processes result in mechani-

cal concentrations of ore minerals or placer deposits. Rocks and ore bodies of primary origin are broken down by disintegration and chemical decomposition (see Chapter 3). The resultant debris is removed from its place of origin by gravity, running water and wind. The minerals preserved in placer deposits must possess three characteristics if they are to survive: high specific gravity, strong resistance to chemical reaction and durability. High specific gravity means that the mineral is relatively heavy and would be transported long distances only by extremely strong currents of water or high winds. Generally the mineral materials would tend to settle in the quieter parts of streams and rivers, and resist the lifting action of wind currents.

Resistance to chemical action is particularly important where the mineral is likely to come into contact with acidulated waters. Durability – hardness, toughness, malleability and cleavage – is essential if the mineral is to survive the rough treatment it receives during the erosion and transportation processes. Durability is determined by the atomic structure of the mineral and by the strength of the bonding between the parts of the atom (see Chapter 1).

Placer deposits can be sub-classified into five groups – fluviatile placers, beach placers, buried placers, eluvial placers and aeolian placers.

Fluviatile placers yield valuable concentrations of minerals ores and are usually easy to work. The transport agent is water and accumulation occurs when the velocity of a stream falls to the level where the mineral cannot be either carried along in suspension or rolled along the bed of the stream.

One of the most common sites for the accumulation of fluviatile placers is on the convex side of a meander, where slack water is found. Since meanders tend to travel across and along the length of a valley, the site of the deposits is constantly changing as the course of the stream changes. The early-formed deposits are gradually covered by other debris, and so the search for deposits must go beyond the shoulders of the present-day stream. Hard bands of rock lying across the course of a stream provide barriers against which minerals can accumulate, and potholes form traps in which rich deposits sometimes occur. The velocity of the water often falls at the point where a tributary enters a main river, and these areas are also likely areas of mineral deposition.

Gold placers are the earliest form of fluviatile deposits mentioned in legend or literature. Early accounts mention the area of Colchis at the eastern end of the Black Sea where rich harvests of gold were found in placer deposits formed by rivers flowing from the Caucasus Mountains. To recover the gold a fleece was wetted and used to line a hollowed-out tree trunk. The silt-laden waters passed across the fleece and the gold adhered to it. The dross was washed away and the larger fragments of gold picked off. The fleece was then dried and beaten to recover the finer gold particles.

RIGHT **Broken Hill, New South Wales, Australia** Part of the surface workings at Rio Tinto Zinc's lead, zinc and silver mines at Broken Hill.

BELOW **Diamond mines, South Africa** Many diamonds are obtained from placer deposits, which, generally speaking, are easier to work than diamond mines. Here the mother lode is being mined below ground in difficult and sometimes dangerous conditions.

This may well have been the origin of the legend of the Golden Fleece which Jason and the Argonauts set out to find. Although as a method of gold recovery it is very primitive, it was still being used in the 1930s in areas of Brazil, where cowhides were used instead of fleeces.

Every prospector's dream is to find a nugget – or chunk – of gold. The largest nugget ever found was the Welcome Stranger found at Ballarat in Australia, which weighed 2,280 ounces. But most placer gold occurs in tiny flat pellets or as fine dust known as 'colours'. Gold placers probably had their sources in rich 'mother lodes', and prospectors often followed the placer deposits upstream in an attempt to discover them. Sometimes they were successful, as in the discovery of

LEFT **Tin mining, Malaya** Water hoses are directed at the ore-bearing rocks. The pressure of the water jet breaks up the rocks. Tin is the heaviest material and is left behind when all the other sediment is washed away. The tin ore is then refined and the metal extracted from it. In our everyday 'tins' there is only a thin covering of tin on the base metal.

ABOVE **Mining, Broken Hill, New South Wales, Australia** Mining an ore-rich face below ground. When it is refined, the rock will yield silver, lead and zinc. The difficult and dangerous conditions in which the miners work can be seen from the photograph. The miner on the right is wearing ear guards to protect his hearing from the noise of the machinery.

ABOVE **Diamond mining in South Africa** The diamond-bearing earth is gathered up by these huge machines. It may take an average of twenty tons of rock to produce four carats of diamonds.

RIGHT **Fired bricks** 'Biscuits' made of Bovey Tracey ballclay used for domestic fireplaces (Devon fires).

the mother lode countries in California and Australia. Other lodes were never found, as in the great Klondyke gold rush of 1897.

The first placer diamonds seem to have come from India. Graphic descriptions of the mining areas were given by the French jeweller, Tavernier, who visited the Indian fields in the first half of the seventeenth century. His descriptions show that most of the diamonds came from sandy alluvial deposits such as those at Golconda. The vast wealth of the Indian princes can be traced to these alluvial sources. Placer deposits of diamonds were discovered in 1772 in Minas Gerais in Brazil, and in 1844 in Bahia, also in Brazil. The first placer diamonds in South Africa were discovered in 1908.

Other well-known gem placer deposits are those of Sri Lanka where a wide variety of gemstones are found – alexandrite, sapphire, ruby, beryl, tourmaline and many others (see Chapter 6). Madagascar also has important placer gemstone deposits. In Malaysia and Indonesia tin occurs in extensive placers derived from decomposed granitic rocks. Platinum placers are found in the Ural Mountains and in South Africa, Tasmania and New South Wales in Australia, and these were probably derived from basic and from ultra basic rocks such as peridotite and dunite.

Beach placers are often associated with fluviatile placers and in some instances the material is actually derived from such placers. A second source of mineral ore supply is the rocks on seashore areas which break down under the combined action of waves and wind. Onshore wave action and long shore drift affect the concentration of durable minerals with high specific gravities. The diamond-bearing gravels of South West Africa were formed in this way, as were the gold beach placers of Nome, Alaska; the zircon and monazite sands of Travancore, India, and the magnetite placers of Oregon in the United States, New Zealand and Japan.

Buried placers fall into two groups: those deposits buried below the surface by younger rocks, volcanic ash or volcanic lava; and those formed in the same manner but which have been subsequently lifted and lie at higher levels. Most buried placers are gold placers. Those belonging to the first group have often been mined at depth, while erosion sometimes exposes gold placers belonging to the second group.

Eluvial placers occur where deep weathering has taken place in hilly country and the resultant debris has moved downhill under the influence of gravity. The tin placers of Malaysia are good examples of this type of placer deposit. Gold is also found in eluvial placers.

Aeolian placers are comparatively rare and form as a result of the winnowing of light material by winds. Gold is found in Australia, Mexico and southern California in this type of placer.

Non-metallic minerals and rocks

Ore minerals and gemstones are very important, but non-metallic minerals are just as useful and valuable. They include the clays, sands and gravels which are so useful in the construction industries ; common salt; and a variety of useful building materials including granite, slate, marble, sandstone, limestone and flints.

One of the most important basic materials is clay, of

ABOVE **Penrhyn slate quarry, Bethesda, North Wales** Slates ready for shipping. Note the huge piles of waste material in the background. Waste such as this is a feature of slate areas.

LEFT **Flint and sandstone wall, Amesbury, Wiltshire, England** Contrast the surface texture and appearance of flint and sandstone (both dressed) and the cement.

which there are many types. Clay is fine-grained sedimentary material formed either 'in situ' as a direct product of weathering – these are the residual clays – or as an accumulation of the finer products of weathering which have been transported by water, wind or ice and deposited in layers – these are the transported clays. Clays are formed mainly of the clay minerals, the hydrous silicate of aluminium, although they contain varying amounts of other minerals such as lime, magnesium, sodium, potassium and titanium as well as mica and quartz fragments. All the component grains are very small, around 0.002mm maximum diameter. Physically the clays are soft and plastic when wet, and hard and brittle when dry. Among the many useful and valuable products derived from the various clays are china, earthenware, insulating materials and building bricks.

Natural building materials are generally bulky and for their excavation to be economically viable, must be obtained as near the construction sites as possible. The result of this has been the utilization of the best *available* natural mineral rather than the most suitable, which has led to a wide variety of mineral and rock types being used in building work. For example, Aberdeen has many buildings constructed from locally quarried grey granite; many towns in Yorkshire and Lincolnshire are built mainly from red sandstone quarried in the nearby Pennines; the Penant sandstone

of South Wales is widely used throughout the valley towns; and the resistant flints which are underlain by chalk provide building stone in parts of Sussex. On the other hand, some stone is so highly regarded for its properties of beauty and resilience that it is sought in more widespread areas. For example, the granites from Shap Fell, Westmoreland and Peterhead in Scotland are to be seen as facing stone on banks and public buildings all over Britain; the milky-blue larvikite (a syenite) from Norway is used as a facing material throughout Europe; and the orbicular granites from Finland are used in the steps of the railway station of Brussels in Belgium. Before the advent of cheap and abundant building bricks and concrete, the building stones of many towns thus reflected the geology of the area.

Cement is another important building material. The

ABOVE **Aberdeen** Aberdeen is known as the 'granite city' because of the predominance of granite used in its buildings.

LEFT **Copper** Copper is rarely found in its natural pure state, as here, but generally occurs as a universal compound. When exposed to the air it tarnishes into copper oxide.

BELOW **China clay tips, St Stephen, Cornwall** China clay is a pure form of clay that is used in the ceramic industry.

ABOVE **Steelworks at Port Kembla, Australia** Tapping an open hearth furnace at the steelworks of Australian Iron and Steel Ltd.

RIGHT **Pouring billets at Port Kembla** The formation of steel by melting sedimentary iron ore is a similar process to the escaping of magma from the earth's crust.

Romans made a type of cement using lime and volcanic ash. In 1824 an Englishman named Aspdin discovered a new way to make cement. His product looked rather like Portland stone when set, and so became known as Portland cement. Since then there have been enormous advances in cement technology, but all modern cements are made from natural minerals. Cement consists, generally speaking, of 75% limestone and 25% clay. The whole material is sintered, ground and treated and ends up as a grey flour-like material. Other minerals may be added to impart a specific quality – for example, the addition of the mineral gypsum speeds up the setting rate.

Sands and gravels together with crushed and graded stones are the other important ingredients of concrete. The sands and gravels are accumulated under fluviatile conditions (see page 118) and occur either as river gravels or as outwash fans from mountain tracts. Sand also provides the raw material for glass-making, for moulding processes and for the production of

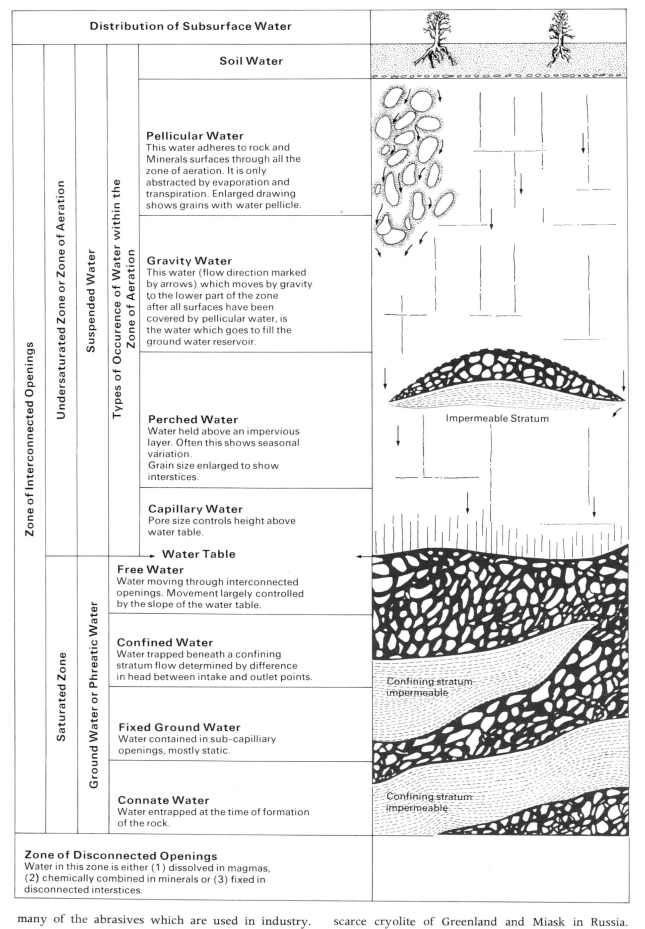

Distribution of Subsurface Water

Soil Water

Pellicular Water
This water adheres to rock and Minerals surfaces through all the zone of aeration. It is only abstracted by evaporation and transpiration. Enlarged drawing shows grains with water pellicle.

Gravity Water
This water (flow direction marked by arrows) which moves by gravity to the lower part of the zone after all surfaces have been covered by pellicular water, is the water which goes to fill the ground water reservoir.

Perched Water
Water held above an impervious layer. Often this shows seasonal variation.
Grain size enlarged to show interstices.

Capillary Water
Pore size controls height above water table.

→ Water Table

Free Water
Water moving through interconnected openings. Movement largely controlled by the slope of the water table.

Confined Water
Water trapped beneath a confining stratum flow determined by difference in head between intake and outlet points.

Fixed Ground Water
Water contained in sub-capillary openings, mostly static.

Connate Water
Water entrapped at the time of formation of the rock.

Zone of Disconnected Openings
Water in this zone is either (1) dissolved in magmas, (2) chemically combined in minerals or (3) fixed in disconnected interstices.

Left-axis labels: Zone of Interconnected Openings; Undersaturated Zone or Zone of Aeration; Suspended Water; Types of Occurence of Water within the Zone of Aeration; Saturated Zone; Ground Water or Phreatic Water

In diagram: Impermeable Stratum; Confining stratum impermeable; Confining stratum impermeable

LEFT **Distribution of the subsurface waters** The division into the zone of aeration or zone of oxidation and the saturated zone is important in mineralization. It is here that chemical change takes place.

RIGHT **Geological conditions giving rise to springs** Notice the importance of rock texture and structure which together determine their porosity and permeability.
1 Water infiltrating through joints and cleavage in shale. Water is lost into the joints in the limestone unless the shale/limestone junction is the plane of saturation (SP).
2 The fault brings shale (impervious) against sandstone (pervious). Water banks up in the sandstone and overflows at SP.
3 Solution channels in limestone allow an almost free flow from the recharge surface to spring SP. The impervious shale below the limestone deflects water at SP.
4 Groundwater may accumulate in the joints in igneous rocks. In this case impervious glacial clay seals the exits and the water table rises to give the spring SP by overflow.
5 The igneous dykes provide an impervious barrier and springs develop by overflow.
6 Impervious layers of shale hold up perched water tables. These often give springs with a seasonal flow.

FAR RIGHT **The Irton Well, Irton, Yorkshire** (after Walters: *The Nation's Water Supply*) A geological section shows the effect of faulting which has brought impervious rocks together. The impervious blanket of boulder clay prevents surface inflow.

RIGHT **Artesian well** This diagram shows how the conditions necessary for the formation of an artesian well are developed in a syncline.

---

many of the abrasives which are used in industry.

There are many other minerals of commercial importance, of course, among them rock salt, which is used in cooking and food preservation and in the manufacture of explosives, glass and soap. Other minerals of commercial importance include fluorite, used as a flux in the metallurgical industry, and the scarce cryolite of Greenland and Miask in Russia. Until a way of synthesizing this mineral was found, its excavation was essential to the production of aluminium. Many minerals are important additives to steel. For example, vanadium is used in the manufacture of special steels; chrome is used in stainless steel manufacture; tungsten is used for making very

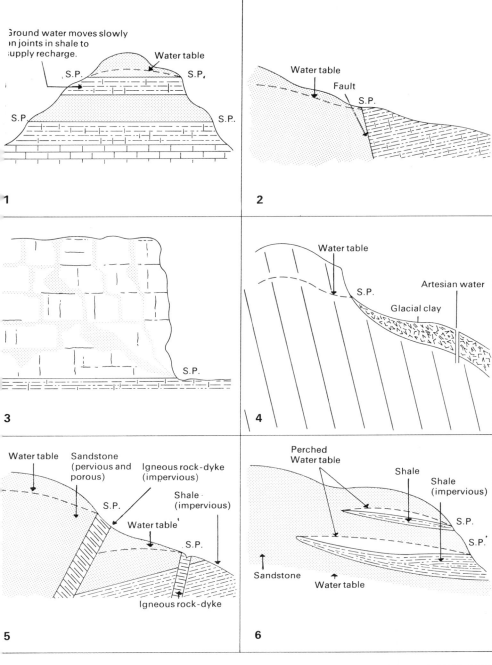

1

Ground water moves slowly on joints in shale to supply recharge.

Water table

S.P. S.P.

S.P. S.P.

2

Water table

Fault

S.P.

3

S.P.

4

Water table

S.P.

Artesian water

Glacial clay

5

Water table Sandstone (pervious and porous) Igneous rock-dyke (impervious)

S.P.

Shale (impervious)

Water table

S.P.

Igneous rock-dyke

6

Perched Water table

Shale

Shale (impervious)

S.P.

S.P.

Sandstone

Water table

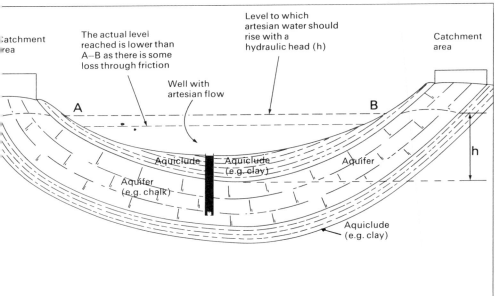

Catchment area

The actual level reached is lower than A–B as there is some loss through friction

Well with artesian flow

Level to which artesian water should rise with a hydraulic head (h)

Catchment area

A

B

h

Aquiclude Aquiclude (e.g. clay) Aquifer

Aquifer (e.g. chalk)

Aquiclude (e.g. clay)

hard steels for cutting and drilling machinery. The pitchblende group of minerals from Canada, Central Africa, and Czechoslovakia are necessary minerals in the medical field and in the production of nuclear energy.

The fuels

The natural fuels are coal, oil and natural gas. Coal is a sedimentary rock composed mainly of accumulated plant material (see Chapter 3). Scientists think coal deposits formed from thick layers of decaying vegetation, which were covered with layers of other materials, and slowly became compacted and solidified. It takes about 16 ft of plant matter for one ft of coal. Peat, used as a fuel in some areas, is formed in a similar way. So too is lignite, which is a soft brown coal that has not undergone the whole coal formation process. Natural gas and crude oil have much the same origins as coal. Both are recovered in much the same way and they are often found together. They are formed by the accumulation of microscopic plants and animals, probably on the sea-bed. After burial this accumulate becomes trapped in hollows in the earth's crust. The most suitable areas for trapping oil and gas occur where the rocks have folded into an arch structure known as an anticline. These features are clues to geologists searching for oil deposits. The oil or natural gas is extracted by drilling. Oil needs extensive preparation and refining

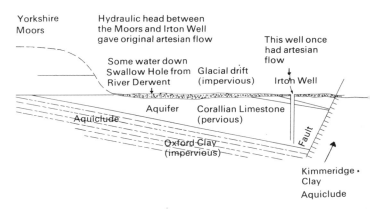

Yorkshire Moors

Hydraulic head between the Moors and Irton Well gave original artesian flow

Some water down Swallow Hole from River Derwent

Glacial drift (impervious)

This well once had artesian flow

Irton Well

Aquifer Corallian Limestone (pervious)

Aquiclude

Oxford Clay (impervious)

Fault

Kimmeridge Clay Aquiclude

before it is ready for use, but natural gas needs little refining and can be pumped directly to the consumer.

Water supplies

Water is essential to all life and industrial activity. It is present on the surface of the earth as streams, rivers, lakes and oceans, and also below the surface in natural reservoirs. Natural underground reservoirs occur mainly in sedimentary rocks in areas where there is a watertight sealing layer of underlying rock – unfractured igneous or metamorphic rocks, or fine-grained sedimentary rocks, for example. Natural springs occur where the junction of the reservoir rock with the sealing rock is exposed above ground. Areas of watertight rock can also be sealed off by means of dams to provide man-made reservoirs.

The minerals and rocks mined commercially by man are all natural in their origin. They are wasting assets for no replacement is taking place, so far as we know, and there is fear among scientists and conservationists that supplies of these natural resources will one day be exhausted.

Acknowledgments

The Publishers would like to thank W. Gwynne Morris and Tony Burrett, Jean Atcheson and Tressilian Nicholas for their help in the preparation of this book.

The publishers would like to thank Barrington Barber/Augustine Studios for drawing the maps.

The publishers would like to thank the authorities and trustees of the following collections and museums for their kind permission to reproduce the illustrations in this book:

Aerofilms 45 below left
Almenna Bókafélagid 38–9
Australian Information Service 105 above, 107 below, 116 below left 123 both
Barnaby's Picture Library 54, 63 above, 70 both, 86, 105 below left, 110–11, 113
Bavaria Verlag 5 centre, 43 centre, 87 above right
John Beecham 80 below
Camera Press 16–17, 103 both
C & D Cannon 67 right
J Allan Cash 6–7, 73 below
Robert Chew 82–3
Daily Telegraph Picture Library 104 below
H. S. Davidson 55 left
De Beers 90–1, 93 below, 96 all, 97 all, 100 below, 101 both, 118, 120 above
European Colour Library/Carlo Bevilacqua 30 left, 31 above, 76–7, 80 above, 94 right, 98 right, 115 above left, 116 below right
Garrard's London 104 above
Institute of Geological Sciences, London 2–3, 4 top right, 5 top, 9 left, 10 both, 11 both, 22, 22–3, 27 right, 28, 29 below left, 31 below, 34 both, 34–5, 35 all, 41 both, 47 both, 52, 53 all, 55 above and below right, 56, 57 all, 60, 61 all, 62 right, 64 above, 65 both, 68 above, 69 above, 72 above, 74–5, 76 above, 77, 79 both, 81 both, 84 centre right, 85, 88 both, 89 both, 95 above, 99 below, 105 below right, 106, 107 above, 112 below, 112–13, 120 below
Dr Gilbert Kelling 64 below
NASA 66
National Coal Board 112 above
Natural Science Photos 76 left, 78–9
J Oliver 67 left
Picturepoint 14 above, 42 below, 43 above, 87 left, 100 centre, 102, 115 right, 119 below left and right
Popperfoto 117
Rio Tinto Zinc 5 below, 30 right, 115 below left, 119 above left
Brian Simpson 1, 33, 45 below right, 50 below right, 51, 63 below, 121 above, 122 below right
Ian Simpson 69 below
Spectrum 122 above right
R. T. Way 8, 8–9, 16 left, 40, 44 below, 45 above left and right, 46 left, 46–7, 50–1, 66–7, 68 below, 72 below, 84 centre left, 121 below
Zefa 4 left and below right, 14 centre and below, 18–19, 23 both, 26, 27 left, 29 above and below right, 32 above, 42–3, 43 below 44 above, 50 below left, 58–9, 62 left, 71, 78, 87 below right, 94 left, 95 below, 98 left, 99 above, 115, 122 left
Jacket illustrations: (front and back) Minerva Terrace Yellowstone National Park (Zefa); (front flap) amethyst crystals (Zefa); (back flap) Brazilian tourmaline (Zefa); (endpapers) fault (Institute of Geological Sciences).

Index